Selected Titles in This Series

707 **Henning Krause,** The spectrum of a module category, 2001

706 **Jonathan Brundan, Richard Dipper, and Alexander Kleshchev,** Quantum linear groups and representations of $GL_n(\mathbb{F}_q)$, 2001

705 **I. Moerdijk and J. J. C. Vermeulen,** Proper maps of toposes, 2000

704 **Jeff Hooper, Victor Snaith, and Min van Tran,** The second Chinburg conjecture for quaternion fields, 2000

703 **Erik Guentner, Nigel Higson, and Jody Trout,** Equivariant E-theory for C^*-algebras, 2000

702 **Ilijas Farah,** Analytic guotients: Theory of liftings for quotients over analytic ideals on the integers, 2000

701 **Paul Selick and Jie Wu,** On natural coalgebra decompositions of tensor algebras and loop suspensions, 2000

700 **Vicente Cortés,** A new construction of homogeneous quaternionic manifolds and related geometric structures, 2000

699 **Alexander Fel′shtyn,** Dynamical zeta functions, Nielsen theory and Reidemeister torsion, 2000

698 **Andrew R. Kustin,** Complexes associated to two vectors and a rectangular matrix, 2000

697 **Deguang Han and David R. Larson,** Frames, bases and group representations, 2000

696 **Donald J. Estep, Mats G. Larson, and Roy D. Williams,** Estimating the error of numerical solutions of systems of reaction-diffusion equations, 2000

695 **Vitaly Bergelson and Randall McCutcheon,** An ergodic IP polynomial Szemerédi theorem, 2000

694 **Alberto Bressan, Graziano Crasta, and Benedetto Piccoli,** Well-posedness of the Cauchy problem for $n \times n$ systems of conservation laws, 2000

693 **Doug Pickrell,** Invariant measures for unitary groups associated to Kac-Moody Lie algebras, 2000

692 **Mara D. Neusel,** Inverse invariant theory and Steenrod operations, 2000

691 **Bruce Hughes and Stratos Prassidis,** Control and relaxation over the circle, 2000

690 **Robert Rumely, Chi Fong Lau, and Robert Varley,** Existence of the sectional capacity, 2000

689 **M. A. Dickmann and F. Miraglia,** Special groups: Boolean-theoretic methods in the theory of quadratic forms, 2000

688 **Piotr Hajłasz and Pekka Koskela,** Sobolev met Poincaré, 2000

687 **Guy David and Stephen Semmes,** Uniform rectifiability and quasiminimizing sets of arbitrary codimension, 2000

686 **L. Gaunce Lewis, Jr.,** Splitting theorems for certain equivariant spectra, 2000

685 **Jean-Luc Joly, Guy Metivier, and Jeffrey Rauch,** Caustics for dissipative semilinear oscillations, 2000

684 **Harvey I. Blau, Bangteng Xu, Z. Arad, E. Fisman, V. Miloslavsky, and M. Muzychuk,** Homogeneous integral table algebras of degree three: A trilogy, 2000

683 **Serge Bouc,** Non-additive exact functors and tensor induction for Mackey functors, 2000

682 **Martin Majewski,** ational homotopical models and uniqueness, 2000

681 **David P. Blecher, Paul S. Muhly, and Vern I. Paulsen,** Categories of operator modules (Morita equivalence and projective modules, 2000

680 **Joachim Zacharias,** Continuous tensor products and Arveson's spectral C^*-algebras, 2000

679 **Y. A. Abramovich and A. K. Kitover,** Inverses of disjointness preserving operators, 2000

(*Continued in the back of this publication*)

The Spectrum of a Module Category

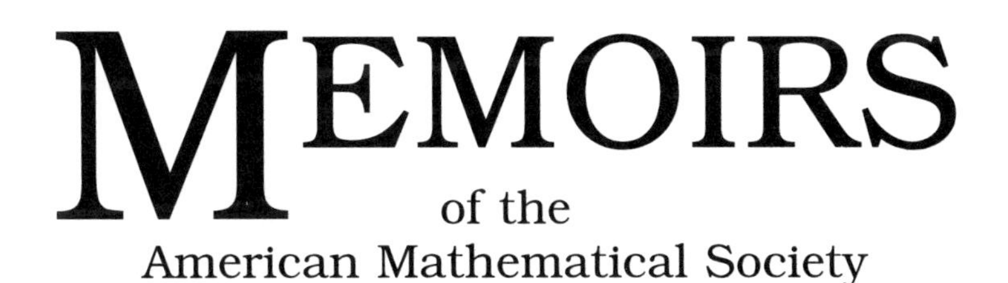

Number 707

The Spectrum of a Module Category

Henning Krause

January 2001 • Volume 149 • Number 707 (second of 4 numbers) • ISSN 0065-9266

American Mathematical Society
Providence, Rhode Island

1991 *Mathematics Subject Classification.*
Primary 16D70, 16D90, 16G60; Secondary 16G70, 16P70, 16S90, 18G25.

Library of Congress Cataloging-in-Publication Data

Krause, Henning, 1962–
The spectrum of a module category / Henning Krause.
p. cm. — (Memoirs of the American Mathematical Society, ISSN 0065-9266 ; no. 707)
"January 2001, Volume 149, Number 707 (second of 4 numbers)."
Includes bibliographical references.
ISBN 0-8218-2618-2 (alk. paper)
1. Injective modules (Algebra) 2. Categories (Mathematics) I. Title. II. Series.
QA3 .A57 no. 707
[QA247.3]
510 s—dc21
[512′.74] 00-046916

Memoirs of the American Mathematical Society

This journal is devoted entirely to research in pure and applied mathematics.

Subscription information. The 2001 subscription begins with volume 149 and consists of six mailings, each containing one or more numbers. Subscription prices for 2001 are $494 list, $395 institutional member. A late charge of 10% of the subscription price will be imposed on orders received from nonmembers after January 1 of the subscription year. Subscribers outside the United States and India must pay a postage surcharge of $31; subscribers in India must pay a postage surcharge of $43. Expedited delivery to destinations in North America $35; elsewhere $130. Each number may be ordered separately; *please specify number* when ordering an individual number. For prices and titles of recently released numbers, see the New Publications sections of the *Notices of the American Mathematical Society.*

Back number information. For back issues see the *AMS Catalog of Publications.*

Subscriptions and orders should be addressed to the American Mathematical Society, P. O. Box 845904, Boston, MA 02284-5904. *All orders must be accompanied by payment.* Other correspondence should be addressed to Box 6248, Providence, RI 02940-6248.

Memoirs of the American Mathematical Society is published bimonthly (each volume consisting usually of more than one number) by the American Mathematical Society at 201 Charles Street, Providence, RI 02904-2294. Periodicals postage paid at Providence, RI. Postmaster: Send address changes to Memoirs, American Mathematical Society, P. O. Box 6248, Providence, RI 02940-6248.

This publication is indexed in *Science Citation Index*®, *SciSearch*®, *Research Alert*®, *CompuMath Citation Index*®, *Current Contents*®*/Physical, Chemical & Earth Sciences.*
Printed in the United States of America.

♾ The paper used in this book is acid-free and falls within the guidelines established to ensure permanence and durability.
Visit the AMS home page at URL: `http://www.ams.org/`

10 9 8 7 6 5 4 3 2 1 06 05 04 03 02 01

Contents

Abstract

These notes present an introduction into the spectrum of the category of modules over a ring. We discuss the general theory of pure-injective modules and concentrate on the isomorphism classes of indecomposable pure-injective modules which form the underlying set of this spectrum. The interplay between the spectrum and the category of finitely presented modules provides new insight into the geometrical and homological properties of the category of finitely presented modules. Various applications from representation theory of finite dimensional algebras are included.

Received by the editor September 21, 1998.

2000 *Mathematics Subject Classification.* Primary 16D70, 16D90, 16G60; Secondary 16G70, 16P70, 16S90, 18G25.

Key words and phrases. Ziegler spectrum, pure injective module, endofinite module, generic module, definable subcategory, coherent functor, Krull-Gabriel dimension, duality, Jacobson radical, tame algebra.

CHAPTER 0

Introduction

LADY S: And what have you been writing about this morning, Mr. K?
K: On the usual subject, Lady S. On Purity.
LADY S: That must be such a very, very interesting thing to write about.
K: It is the one subject of really national importance, nowadays, Lady S.
OSCAR WILDE, from: A woman of no importance.

Let R be a ring (associative with identity) and denote by $\operatorname{Mod} R$ the category of (right) R-modules. The aim of these notes is to present an introduction into the spectrum of the category $\operatorname{Mod} R$ and the machinery which is related to it.

The motivation for this work comes from the representation theory of finite dimensional algebras. If R is a finite dimensional algebra, then one is usually interested in geometrical and homological properties of the category $\operatorname{mod} R$ of finitely generated R-modules. However, in recent years several mathematicians found natural interpretations of concepts from modern representation theory like "representation type" [**17**] or "complexity and varieties" [**10, 69**], using explicitly non-finitely generated modules. This process of passing from $\operatorname{mod} R$ to the category $\operatorname{Mod} R$ of all R-modules (a change of paradigm [**56**]) inevitably involves new concepts and techniques, and it is one of our principal aims to present some of them.

Most important for the applications in representation theory is the concept of "purity" because the pure-injective modules play a prominent role among the non-finitely generated modules. Recall that a module M over some k-algebra R (k a commutative field) is pure-injective if and only if the canonical R-linear map $M \to M^{**}$ splits ($M^* = \operatorname{Hom}_k(M, k)$ denotes the usual k-dual). Although the language of R-modules suffices to give this definition we shall adopt the following more abstract approach. There exists an abelian Grothendieck category $D(R)$ and a fully faithful functor $\operatorname{Mod} R \to D(R)$ which identifies the pure-injective R-modules with the injective objects in $D(R)$. Using this embedding we can derive various results about pure-injective modules from the existing theory for injective objects in Grothendieck categories. The isomorphism classes of indecomposable pure-injective R-modules are of particular interest. They form the underlying set, denoted by $\operatorname{Ind} R$, of the spectrum of $\operatorname{Mod} R$. We shall consider two topologies on $\operatorname{Ind} R$: the Ziegler topology which Ziegler introduced in model-theoretic terms [**80**], and the Zariski topology which reflects the geometric nature of this spectrum [**26**].

To exhibit the interplay between the spectrum $\operatorname{Ind} R$ and the category $\operatorname{mod} R$ of finitely generated modules is the main theme of this work. In order to illustrate these ideas let us explain one of our results. Viewing the category $\operatorname{mod} R$ as a ring with several objects, we introduce a new class of ideals in $\operatorname{mod} R$ which we call fp-idempotent. It is well-known that any ideal $\mathfrak{I}$ in $\operatorname{mod} R$ is idempotent if and only if the additive functors $\operatorname{mod} R \to \operatorname{Ab}$ into the category of abelian groups which vanish on $\mathfrak{I}$ are closed under forming extensions. Therefore we call an ideal $\mathfrak{I}$ fp-idempotent if the finitely presented functors $\operatorname{mod} R \to \operatorname{Ab}$ vanishing on $\mathfrak{I}$ are closed under forming extensions, and we establish an inclusion preserving bijection between these ideals and the Ziegler-closed subsets of $\operatorname{Ind} R$. From this we deduce a new characterization of algebras having tame representation type. In fact, we prove that a finite dimensional algebra over some algebraically closed field has tame representation type if and only if for every $n \in \mathbb{N}$ there are only finitely many fp-idempotent and nilpotent ideals in $\operatorname{mod} R$ which are contained in the ideal generated by the identity maps of the indecomposable R-modules of dimension n. Recall that an algebra R is of tame representation type if for every $n \in \mathbb{N}$ the indecomposable R-modules of dimension n belong to a finite number of continuous 1-parameter families. Note that our characterization of tameness involves only finitely generated R-modules; it provides therefore an answer to a question raised by Ringel [**72**, p.144]. The proof of this result uses the machinery which is related to the spectrum $\operatorname{Ind} R$, and we hope to convince the reader that once one accepts the idea of introducing infinitely generated modules, one obtains new insight into the structure of the category of finitely generated modules.

Modern representation theory of finite dimensional algebras is based to a large extent on the functorial approach of Auslander and Reiten. In these notes we keep the functorial point of view which Auslander and Reiten initiated into representation theory, but we do not restrict ourselves to the finite level of classical Auslander-Reiten theory which we now explain. To this end denote by $\mathfrak{rad}$ the Jacobson radical of the category $\operatorname{mod} R$ of finitely generated R-modules. The problem of classifying the (indecomposable) objects in $\operatorname{mod} R$ is essentially the problem of understanding the semi-simple quotient $\operatorname{mod} R/\mathfrak{rad}$. The most popular concepts from classical Auslander-Reiten theory are the almost split sequences in $\operatorname{mod} R$ and the Auslander-Reiten quiver of R. These concepts are defined in terms of maps which belong to $\mathfrak{rad}^1 \setminus \mathfrak{rad}^2$. Therefore classical Auslander-Reiten theory covers the "finite part" of $\operatorname{mod} R$, i.e. the quotient $\operatorname{mod} R/\mathfrak{rad}^\omega$ where $\mathfrak{rad}^\omega = \bigcap_{n\in\mathbb{N}} \mathfrak{rad}^n$. The algebra R is of finite representation type if and only if $\mathfrak{rad}^\omega = 0$, and therefore it seems to be important to study also the maps which belong to $\mathfrak{rad}^\omega$.

On the functorial level the finiteness condition in the work of Auslander and Reiten can be explained as follows. Consider the Yoneda embedding

$$\operatorname{mod} R \longrightarrow ((\operatorname{mod} R)^{\mathrm{op}}, \operatorname{Ab}), \quad M \mapsto \operatorname{Hom}_R(-, M)$$

of $\operatorname{mod} R$ into the category of contravariant additive functors $\operatorname{mod} R \to \operatorname{Ab}$ where Ab denotes the category of abelian groups. Every right almost split map $M \to N$ gives rise to a presentation $\operatorname{Hom}_R(-, M) \to \operatorname{Hom}_R(-, N) \to S_N \to 0$ of a simple functor in $((\operatorname{mod} R)^{\mathrm{op}}, \operatorname{Ab})$, and it is the crucial observation of Auslander and Reiten that all simple objects in $((\operatorname{mod} R)^{\mathrm{op}}, \operatorname{Ab})$ arise in this way. Therefore classical Auslander-Reiten theory is mainly concerned with the objects of finite length in the functor category $((\operatorname{mod} R)^{\mathrm{op}}, \operatorname{Ab})$. However, there are finitely presented functors of infinite length if the algebra R has infinite representation type. In fact, $\mathfrak{rad}^\omega$ is precisely the ideal of maps in $\operatorname{mod} R$ which are annihilated by all functors $\operatorname{mod} R \to \operatorname{Ab}$ of finite length, and we observe in this context the importance of studying the functors in $((\operatorname{mod} R)^{\mathrm{op}}, \operatorname{Ab})$ which are of infinite length.

In contrast to the classical approach of Auslander and Reiten, we exhibit in this work several phenomena which occur only in the "infinite part" of $\operatorname{mod} R$. In particular, we investigate the structure of $\mathfrak{rad}^{\omega}$ and study functors on $\operatorname{mod} R$ which are not of finite length. We refer to Lenzing's Trondheim lectures for an outline of this program [**58**].

We end this discussion of the Auslander-Reiten approach with another point which seems to be crucial. The work of Auslander and Reiten is based on the Yoneda embedding $\operatorname{mod} R \to ((\operatorname{mod} R)^{\mathrm{op}}, \operatorname{Ab})$ whereas in this work we mainly use the embedding

$$\operatorname{mod} R \longrightarrow (\operatorname{mod} R^{\mathrm{op}}, \operatorname{Ab}), \quad M \mapsto M \otimes_R -$$

via the tensor functors. The Yoneda embedding identifies the finitely generated modules with projective objects in the functor category $((\operatorname{mod} R)^{\mathrm{op}}, \operatorname{Ab})$, and therefore many results about $\operatorname{mod} R$ can be derived from results about projective covers and minimal projective presentations in $((\operatorname{mod} R)^{\mathrm{op}}, \operatorname{Ab})$. However, there exists a projective cover for every object in $((\operatorname{mod} R)^{\mathrm{op}}, \operatorname{Ab})$ if and only if the algebra R is of finite representation type. On the other hand, injective envelopes always exist in $(\operatorname{mod} R^{\mathrm{op}}, \operatorname{Ab})$. Note that $(\operatorname{mod} R^{\mathrm{op}}, \operatorname{Ab})$ and $((\operatorname{mod} R)^{\mathrm{op}}, \operatorname{Ab})$ are equivalent categories since $\operatorname{mod} R^{\mathrm{op}}$ and $(\operatorname{mod} R)^{\mathrm{op}}$ are equivalent, but we prefer to identify R-modules with injective objects in $(\operatorname{mod} R^{\mathrm{op}}, \operatorname{Ab})$ via the tensor embedding since the existence of injective envelopes can be used without any assumptions. Let us mention that the embedding $M \mapsto M \otimes_R -$ already occurs in Auslander's work. In fact, in his Temple notes [**5**] he devotes a few pages to studying pure-injective modules, but 20 years ago this point of view did not receive much attention.

These notes are divided into 15 chapters, each of which is subdivided into sections. We describe now the basic notions and results of this work chapter by chapter. We fix a ring R, denote by $\operatorname{Mod} R$ the category of all R-modules, and $\operatorname{mod} R$ denotes the full subcategory of finitely presented R-modules.

In Chapter 1 we recall the basic facts about the Grothendieck category $D(R)$ of **additive functors** $\operatorname{mod} R^{\mathrm{op}} \to \operatorname{Ab}$ from the category of finitely presented R^{op}-modules to the category of abelian groups. In particular we discuss the embedding $\operatorname{Mod} R \to D(R)$ which sends an R-module M to the tensor functor $M \otimes_R -$. The finitely presented functors in $D(R)$ play a prominent role. They form an abelian category which is denoted by $C(R)$; it is the free abelian category over R (viewed as a category with a single object).

Chapter 2 is devoted to the fundamental concept of a **definable subcategory** of $\operatorname{Mod} R$. This notion has its origin in model theory of modules; in this context a definable subcategory corresponds to a complete theory of modules [**61, 80**]. However, definable subcategories were introduced formally by Crawley-Boevey when their relevance became apparent in representation theory [**19**]. Given a collection Φ of maps in $\operatorname{Mod} R$ we say that an R-module M is Φ-injective if every map $\phi\colon X \to Y$ in Φ induces a surjection $\operatorname{Hom}_R(Y, M) \to \operatorname{Hom}_R(X, M)$. We denote by $(\operatorname{Mod} R)_{\Phi}$ the full subcategory of Φ-injective R-modules. A subcategory $\mathcal{X}$ of $\operatorname{Mod} R$ is called definable if $\mathcal{X} = (\operatorname{Mod} R)_{\Phi}$ for some collection Φ of maps between finitely presented R-modules. Such categories are automatically closed under taking direct limits, products, and pure submodules in $\operatorname{Mod} R$. Using some model-theoretic arguments it has been shown by Crawley-Boevey that these properties characterize the definable subcategories of $\operatorname{Mod} R$. We give a new proof of this fact, using the localization theory for locally coherent Grothendieck categories which is explained in Appendix A. Next we exhibit the **Ziegler spectrum** of R. This is, by definition, the set $\operatorname{Ind} R$ of isomorphism classes of indecomposable pure-injective R-modules, together with a topology which Ziegler introduced in model-theoretic terms [**80**]. The closed subsets are precisely those of the form $\mathbf{U}_\Phi = \operatorname{Ind} R \cap (\operatorname{Mod} R)_\Phi$ for some Φ in $\operatorname{mod} R$, and we prove that each definable subcategory is completely determined by the corresponding Ziegler-closed set.

THEOREM. *An R-module M belongs to a definable subcategory* $(\operatorname{Mod} R)_\Phi$ *if and only if M is isomorphic to a pure submodule of some product of modules in* $\mathbf{U}_\Phi$.

There is also the concept of a definable quotient category of $\operatorname{Mod} R$ which is defined with respect to a definable subcategory [**52, 49**]. We discuss the basic properties of a definable quotient category and conclude this chapter with a long list of examples.

In Chapter 3 we discuss two types of **left approximations**. The first type are the **left almost split maps** which have been introduced by Auslander and Reiten [**6**]. This concept is one of the most successful in modern algebra representation theory. Here we give a new characterization of the existence of a left almost split map starting in a pure-injective R-module, using only the modules in $\operatorname{Ind} R$.

THEOREM. *There exists a left almost split map $M \to M'$ starting in a pure-injective R-module M if and only if for every product $N = \prod_{i\in I} N_i$ of modules in* $\operatorname{Ind} R$ *such that M is isomorphic to a direct summand of N there exists $i \in I$ such that $M \simeq N_i$.*

Also we discuss the connection between left almost split maps in $\operatorname{mod} R$ and $\operatorname{Mod} R$. The second type of left approximations is defined with respect to a class $\mathcal{C}$ of R-modules. Following [**7**], $\mathcal{C}$ is said to be **covariantly finite** if for every R-module M there exists a left $\mathcal{C}$-approximation $M \to N$, i.e. N belongs to $\mathcal{C}$ and the induced map $\operatorname{Hom}_R(N, C) \to \operatorname{Hom}_R(M, C)$ is surjective for all C in $\mathcal{C}$. We prove that every definable subcategory of $\operatorname{Mod} R$ is covariantly finite and give criteria for the existence of **minimal left approximations**. Recall that a map $\phi\colon M \to N$ is left minimal if every endomorphism ψ of N satisfying $\psi \circ \phi = \phi$ is an automorphism. For example, there exists always a minimal left $\mathcal{C}$-approximation for M if there exists a left $\mathcal{C}$-approximation $M \to N$ with pure-injective N. In this way we generalize the concept of a (pure) injective envelope of a module M.

While there is no general **duality** between $\operatorname{Mod} R$ and $\operatorname{Mod} R^{\mathrm{op}}$ there exists nevertheless a relation between certain classes of right and left R-modules. Some of these facts are collected in Chapter 4. The basic tool is the duality $C(R) \to C(R^{\mathrm{op}})$ between the abelian categories of finitely presented functors in $D(R)$ and $D(R^{\mathrm{op}})$. For example, we obtain a bijective correspondence between the definable subcategories of $\operatorname{Mod} R$ and $\operatorname{Mod} R^{\mathrm{op}}$ since any definable subcategory of $\operatorname{Mod} R$ is determined by a Serre subcategory of $C(R)$. Moreover, we introduce a class $\operatorname{Ref} R$ of R-modules which we call **pure-reflexive**, and construct bijections $M \mapsto M^\vee$ between $\operatorname{Ref} R$ and $\operatorname{Ref} R^{\mathrm{op}}$ satisfying $M^{\vee\vee} = M$ for all M. For example, if R is an algebra over a commutative field k, then $M^\vee = M^*$ for every finite dimensional R-module M, whereas $M^\vee$ is a direct summand of M^* if M is of infinite dimension ($M^* = \operatorname{Hom}_k(M,k)$ denotes the usual k-dual). The construction of the map $M \mapsto M^\vee$ provides a new interpretation of Herzog's elementary duality [**39**]. In fact, our approach leads to an axiomatic description of the map $M \mapsto M^\vee$ which is based on the following simple conditions: $(M \coprod N)^\vee = M^\vee \coprod N^\vee$; $M^{\vee\vee} = M$; and $M^\vee$ is a direct summand of M^*.

In Chapter 5 we introduce a new class of ideals which we call **fp-idempotent**. Viewing the category $\operatorname{mod} R$ as a ring with several objects, it is well-known that any ideal $\mathfrak{I}$ in $\operatorname{mod} R$ is idempotent if and only if the additive functors $\operatorname{mod} R \to \mathrm{Ab}$ vanishing on $\mathfrak{I}$ are closed under forming extensions. We say therefore that $\mathfrak{I}$ is fp-idempotent provided that the finitely presented functors $\operatorname{mod} R \to \mathrm{Ab}$ vanishing on $\mathfrak{I}$ are closed under forming extensions. An fp-idempotent ideal $\mathfrak{I}$ is usually not idempotent, but it can be shown that $\mathfrak{I} = \mathfrak{J} \cap \operatorname{mod} R$ for some idempotent ideal $\mathfrak{J}$ in $\operatorname{Mod} R$. More precisely, we denote for every additive subcategory $\mathcal{X}$ of $\operatorname{Mod} R$ by $[\mathcal{X}]$ the ideal of all maps in $\operatorname{mod} R$ which factor through some module in $\mathcal{X}$.

THEOREM. *Let R be an artin algebra. The assignments*

$$\mathcal{X} \mapsto [\mathcal{X}] \quad \textit{and} \quad \mathfrak{I} \mapsto (\operatorname{Mod} R)_{\operatorname{ann}^{-1}\mathfrak{I}}$$

give mutually inverse and inclusion preserving bijections between the definable subcategories of $\operatorname{Mod} R$ *and the fp-idempotent ideals in* $\operatorname{mod} R$.

Here, we denote for every ideal $\mathfrak{I}$ in $\operatorname{mod} R$ by $\operatorname{ann}^{-1}\mathfrak{I}$ the collection of maps $\phi\colon X \to Y$ in $\operatorname{mod} R$ such that every map $X \to Z$ in $\mathfrak{I}$ factors through ϕ. We obtain also an inclusion preserving bijection between the Ziegler-closed subsets of $\operatorname{Ind} R$ and the fp-idempotent ideals of $\operatorname{mod} R$ by sending a Ziegler-closed subset $\mathbf{U}$ to the ideal of all maps in $\operatorname{mod} R$ which factor through a product of modules in $\mathbf{U}$.

Endofinite modules are the main subject of Chapter 6. Recall that the endolength of an R-module M is the length of M when regarded in the natural way as module over $S = \operatorname{End}_R(M)^{\mathrm{op}}$. The modules of finite endolength are called endofinite. For example, if R is a finite dimensional algebra, then every finite dimensional R-module is endofinite. However, there are usually also indecomposable endofinite modules which are infinite dimensional, and it has been shown by Crawley-Boevey that these modules control the representation type of R, i.e. the complexity of the category of finite dimensional R-modules [**17**]. Here we give some categorical characterizations of endofiniteness. For instance we assign to every R-module M its **endocategory** $\mathcal{E}_M$. This is an exact abelian subcategory of $\operatorname{Mod} S$, and we show that M is endofinite if and only if every object in $\mathcal{E}_M$ is of finite length. Also we prove that M is indecomposable and endofinite if and only if the coproducts of copies of M form a definable subcategory of $\operatorname{Mod} R$. This result motivates the following definition. We call a module M **product-complete** if every product of copies of M is a direct summand of a coproduct of copies of M.

THEOREM. *A module M is endofinite if and only if every direct summand of M is product-complete.*

Endofinite modules are pure-injective and therefore we can study the indecomposable ones as points of the Ziegler spectrum. Finally we assign to every ideal in $\operatorname{mod} R$ a length and show that for an artin algebra R the fp-idempotent ideals of length n correspond to the endofinite modules of endolength n.

In Chapter 7 we assign to a ring R its **Krull-Gabriel dimension** KGdim R. This dimension is defined in terms of a filtration

$$0 = C(R)_{-1} \subseteq C(R)_0 \subseteq C(R)_1 \subseteq \dots$$

of the abelian category $C(R)$; it is therefore a finitely presented version of the Krull dimension of the Grothendieck category $D(R)$ which Gabriel [**26**] introduced in terms of a filtration

$$0 = D(R)_{-1} \subseteq D(R)_0 \subseteq D(R)_1 \subseteq \dots$$

of $D(R)$. There is a also a local variant KGdim M for every R-module M which is automatically bounded by KGdim R. For example, KGdim $R = 0$ if and only if R is of finite representation type, i.e. R is right artinian and there are only finitely many isomorphism classes of indecomposable R-modules. Also, KGdim $M = 0$ if and only if M is endofinite. The Krull-Gabriel dimension of R measures the complexity of the category Mod R, and KGdim $R < \infty$ seems to be a reasonable finiteness condition if one is interested in classification results similar to those for rings of finite representation type. For example, we show that every pure-injective module M with KGdim $M < \infty$ is the pure-injective envelope of a coproduct of modules in Ind R. From this follows that, for every ordinal α, the duality map $M \mapsto M^\vee$ induces a bijection between the isomorphism classes of right and left R-modules which are pure-injective and have Krull-Gabriel dimension α.

The filtration

$$\operatorname{mod} R = \mathfrak{rad}^0 \supseteq \mathfrak{rad}^1 \supseteq \mathfrak{rad}^2 \supseteq \dots$$

of the **Jacobson radical** $\mathfrak{rad} = \operatorname{rad}(\operatorname{mod} R)$ of mod R plays an important role in the representation theory of finite dimensional algebras because almost split sequences in mod R and the Aulander-Reiten quiver of R are defined in terms of maps which belong to $\mathfrak{rad}^1 \setminus \mathfrak{rad}^2$. In Chapter 8 we study two filtrations of the Jacobson radical $\mathfrak{rad}$ which extend the usual filtration $(\mathfrak{rad}^n)_{n\in\mathbb{N}}$. Our first aim is to extend the class of **preinjective modules**. In their study of the representation theory of finite-dimensional tensor algebras Dlab and Ringel described certain modules which they called preprojective and preinjective modules [**23**]. Later Auslander and Smalø defined these concepts for arbitrary artin algebras [**7**]. We define a new **radical series** $(\mathfrak{rad}_\alpha)_\alpha$ of the category mod R which is indexed by the ordinals. Using this radical series, we assign to every finitely presented module M a preinjective dimension pidim M. This dimension is finite if and only if the module is preinjective in the sense of Auslander and Smalø, and we prove also that for every ordinal α the Krull-Gabriel dimension of R is bounded by α if and only if pidim $M < \omega(\alpha+1)$ for every M in mod R. Then we compare for every ordinal α the ideal $\mathfrak{rad}_\alpha$ of mod R with the power $\mathfrak{rad}^\alpha$ of the Jacobson radical of mod R which Prest introduced [**64**].

THEOREM. *Let R be an artin algebra of Krull-Gabriel dimension α. Then $\mathfrak{rad}^{\omega\alpha+n} = 0$ for some $n \in \mathbb{N}$.*

Functors from a different module category Mod S to Mod R play an important role in our analysis of the spectrum of Mod R. There is a natural class of functors Mod $S \to$ Mod R which we call **coherent** since their composition with the forgetful functor Mod $R \to$ Ab arises as the cokernel of a map $\operatorname{Hom}_S(Y,-) \to \operatorname{Hom}_S(X,-)$ where X and Y belong to mod S. In fact, we may think of such a functor as a representation of R in the abelian category $C(S^{\mathrm{op}})$. In Chapter 9 we present various characterizations and some of the basic properties of coherent functors. For example, it is shown that they are precisely those functors which commute with direct limits and products. Another characterization can be used to prove that a coherent functor preserves endofiniteness and the Ziegler topology.

A finite dimensional algebra R over some algebraically closed field is said to have **tame representation type** if for every $n \in \mathbb{N}$ the indecomposable R-modules of dimension n belong to a finite number of continuous 1-parameter families. Our aim in Chapter 10 is to present two new definitions of tameness. Both definitions seem to be more natural. The first one is formulated in terms of endofinite modules and behaves well with respect to functors between module categories. The second definition uses fp-idempotent ideals and is therefore entirely formulated in terms of the category of finitely presented modules. The main results of this chapter are as follows. Let $\operatorname{ind}_n R$ denote the isomorphism classes of finitely presented indecomposable R-modules of endolength n.

THEOREM. *Let R be a finite dimensional algebra over some algebraically closed field. Then the following are equivalent:*

(1) *The algebra R is of tame representation type.*
(2) *For every $n \in \mathbb{N}$ the Ziegler closure of $\operatorname{ind}_n R$ contains only finitely many modules which do not belong to $\operatorname{ind}_n R$.*
(3) *For every $n \in \mathbb{N}$ there are only finitely many fp-idempotent and nilpotent ideals in $\operatorname{mod} R$ which are contained in the ideal which is generated by the identity maps of the modules in $\operatorname{ind}_n R$.*

Using previous results about endofinite modules, fp-idempotent ideals, and coherent functors, the proof that any tame algebra has both properties is fairly elementary. Applying similar techniques, one shows that any algebra of wild representation type cannot have these properties. We obtain therefore our characterization of tameness if we apply the Tame and Wild Theorem which asserts that an algebra is either tame or wild but not both [**24, 14, 17**]. The essential concept for this discussion is that of a generic module. These are indecomposable endofinite modules which are not finitely presented [**17, 51**]. We end this chapter with an explicit description of the relation between generic modules and 1-parameter families of finite dimensional modules.

In Chapter 11 we assign to every collection Φ of maps in $\operatorname{mod} R$ a ring homomorphism $f_\Phi\colon R \to R_\Phi$ which is called the **ring of definable scalars** for Φ. This is an invariant of the definable subcategory $(\operatorname{Mod} R)_\Phi$ which was first introduced by Prest [**63**]. We use a universal property to define the ring of definable scalars but present various alternative constructions for $R \to R_\Phi$. More precisely, the restriction functor corresponding to f_Φ induces an equivalence $(\operatorname{Mod} R_\Phi)_\Psi \to (\operatorname{Mod} R)_\Phi$ for some collection Ψ in $\operatorname{mod} R_\Phi$, and every homomorphism $f\colon R \to S$ having this property induces a homomorphism $g\colon S \to R_\Phi$ such that $f_\Phi = g \circ f$. The first construction of f_Φ uses the free abelian category $C(R)$ over R so that f_Φ is induced by the quotient functor with respect to a Serre subcategory of $C(R)$; the second construction uses a **calculus of left fractions**, and our third approach yields R_Φ as the **biendomorphism ring** of $M_\Phi = (N_\Phi)^\kappa$ where N_Φ is the product of all Φ-injective modules in $\operatorname{Ind} R$ and $\kappa = \operatorname{card} N_\Phi$.

THEOREM. *Let $S = \operatorname{End}_R(M_\Phi)^{\mathrm{op}}$. Then the canonical map $R \to \operatorname{End}_S(M_\Phi)$ is the ring of definable scalars for Φ.*

We include a discussion of **epimorphisms** in the category of rings because every epimorphism $f\colon R \to S$ is the ring of definable scalars for the collection Φ of maps ϕ in $\operatorname{mod} R$ such that $\phi \otimes_R S$ has a left inverse in $\operatorname{mod} S$. In this case restriction via f_Φ induces an equivalence $\operatorname{Mod} R_\Phi \to (\operatorname{Mod} R)_\Phi$. In fact, this observation shows that any epimorphism $f\colon R \to S$ is obtained by adjoining universally left inverses to a collection of maps in $\operatorname{mod} R$. In particular, every element in S is of the form $f(\mathbf{x})^- f(\mathbf{r})$ for some elements $\mathbf{r}, \mathbf{x} \in R^n$ and a left inverse $f(\mathbf{x})^- \in S^n$ for $f(\mathbf{x})$.

Chapter 12 is devoted to studying the definable subcategories of $\operatorname{Mod} R$ having the additional property that the inclusion functor has a **left adjoint**. We present various characterizations of such subcategories and study their basic properties. Examples are the full subcategories of **Φ-continuous modules** for a collection Φ of maps in $\operatorname{mod} R$. Recall from [**29**] that an R-module M is ϕ-continuous with respect to a map $\phi\colon X \to Y$ in $\operatorname{Mod} R$ if the induced map $\operatorname{Hom}_R(Y, M) \to \operatorname{Hom}_R(X, M)$ is bijective. We study also the rings of definable scalars for such definable subcategories and interpret them as localizations for suitable collections Φ in $\operatorname{mod} R$. We consider for R two types of localizations: the **definable localization** is obtained by adjoining universally left inverses for every element in Φ, and the **universal localization** is obtained by adjoining universally two-sided inverses for every element in Φ. This generalizes the classical localization concept for a ring because inverting an element $r \in R$ is the same as inverting the map $R \to R$, $x \mapsto rx$, in $\operatorname{mod} R$. Our discussion of definable and universal localization is motivated by Schofield's universal localization [**11, 75**] which has been widely used in representation theory, in particular in connection with the formation of perpendicular categories [**76, 33**].

In the past several attempts have been made to define non-commutative affine schemes in order to develop some framework for non-commutative algebraic geometry. In Chapter 13 we exhibit the geometrical properties of the spectrum $\operatorname{Ind} R$ and obtain in this way a fairly general approach towards non-commutative geometry which covers some classical cases. A **structure sheaf** $\mathcal{O}_R$ on $\operatorname{Ind} R$ and a functor $\operatorname{Mod} R \to \operatorname{Mod} \mathcal{O}_R$, $M \mapsto \widetilde{M}$, form the main ingredients of our approach. This construction can be described as follows. First we introduce for every pair of objects M, N in a locally coherent Grothendieck category $\mathcal{A}$ a **presheaf** $\operatorname{Hom}_{\mathbf{X}}(M, N)$ **of local morphisms** on the Gabriel spectrum $\operatorname{Sp} \mathcal{A}$ of isomorphism classes of indecomposable injective objects in $\mathcal{A}$. For example, if $\mathcal{A}$ is the category of quasi-coherent sheaves on a noetherian scheme, then our definition of $\operatorname{Hom}_{\mathbf{X}}(M, N)$ yields the usual sheaf of local morphisms between two sheaves M and N because $\operatorname{Sp} \mathcal{A}$ can be identified with the underlying space of the noetherian scheme [**26**]. Given a Ziegler closed subset $\mathbf{X}$ of $\operatorname{Ind} R$ we associate to $\mathbf{X}$ a localizing subcategory $\mathcal{T}$ of $D(R)$ and obtain a functor $\operatorname{Mod} R \to D(R)/\mathcal{T}$, $M \mapsto T_M$, which identifies $\mathbf{X}$ with the Gabriel spectrum of the locally coherent category $D(R)/\mathcal{T}$. This identification yields a new topology on $\mathbf{X}$ which we call **Zariski topology** and which is usually different from the Ziegler topology. In this way we obtain a presheaf $R_{\mathbf{X}} = \operatorname{End}_{\mathbf{X}}(T_R)$ of rings on $\mathbf{X} = \operatorname{Sp}(D(R)/\mathcal{T})$. Following Prest [**63**], we call $R_{\mathbf{X}}$ the **presheaf of definable scalars** because there is an alternative construction which uses the rings of definable scalars. More precisely, if $\mathbf{U}$ is a Zariski-open subset of $\mathbf{X} = \mathbf{U}_\Phi$, then

$$\Gamma(\mathbf{U}, R_{\mathbf{X}}) \simeq \varprojlim_{\mathbf{U}_{\Phi\cup\phi} \subseteq \mathbf{U}} R_{\Phi\cup\phi}$$

where ϕ runs through all maps ϕ in $\operatorname{mod} R$. Analogously, we associate to every R-module M the presheaf $M_{\mathbf{X}} = \operatorname{Hom}_{\mathbf{X}}(T_R, T_M)$. Having constructed the presheaves

$R_{\mathbf{X}}$ and $M_{\mathbf{X}}$ one obtains automatically an associated sheaf of rings $\mathcal{O}_{\mathbf{X}}$ and a sheaf of abelian groups $\widetilde{M}_{\mathbf{X}}$ for every R-module M. The sheaf $\mathcal{O}_{\mathbf{X}}$ is determined by its stalks which are rings of definable scalars.

THEOREM. *The stalk of $\mathcal{O}_{\mathbf{X}}$ at $P \in \mathbf{X}$ is the ring of definable scalars R_{Φ_P} for $\Phi_P = \{\phi \in \operatorname{mod} R \mid P \text{ is } \phi\text{-injective}\}$.*

We discuss some basic properties of the structure sheaf $\mathcal{O}_{\mathbf{X}}$ and the functor $\operatorname{Mod} R \to \operatorname{Mod} \mathcal{O}_{\mathbf{X}}$, $M \mapsto \widetilde{M}_{\mathbf{X}}$, but for the applications we have in mind it is important to make the appropriate choice for the Ziegler-closed subset $\mathbf{X}$ of $\operatorname{Ind} R$. This is the contents of the following two chapters.

In Chapter 14 we investigate the sheaf of definable scalars for a **tame hereditary artin algebra** R. Restricting to an appropriate Ziegler-closed subset $\mathbf{X}$ of $\operatorname{Ind} R$ it is shown that the sheaf $\mathcal{O}_{\mathbf{X}}$ yields a **curve** which parametrizes the finitely generated indecomposable R-modules. This curve has been studied by Baer, Geigle, and Lenzing [**57, 9**], and its centre by Crawley-Boevey [**16**]. The reinterpretation of this parametrizing curve via rings of definable scalars was suggested by Prest [**65**]. Our main result is that the presheaf of definable scalars is actually a sheaf. To be precise let $\mathbf{I} \subseteq \operatorname{Ind} R$ be the set of finitely generated preinjective indecomposables, and denote by $\mathbf{X} = \bar{\mathbf{I}} \setminus \mathbf{I}$ the Ziegler-closed set of all non-finitely generated indecomposables belonging to the Ziegler closure of $\mathbf{I}$. We consider the Zariski topology on $\mathbf{X}$.

THEOREM. *The presheaf of definable scalars $R_{\mathbf{X}}$ is a sheaf, i.e. $R_{\mathbf{X}} = \mathcal{O}_{\mathbf{X}}$. The ring of global sections $\Gamma(\mathbf{X}, \mathcal{O}_{\mathbf{X}})$ is precisely R and the underlying space $\mathbf{X}$ has dimension 1 and contains a unique generic point which is precisely the unique generic R-module.*

For our proof it is not essential that R is hereditary. In fact, the crucial property for this result is the fact that R has Krull-Gabriel dimension 2.

The **Gabriel spectrum** $\mathbf{X} = \operatorname{Sp} R$ of a **right coherent ring** R is another example of a Ziegler-closed subset of $\operatorname{Ind} R$ which we study in Chapter 15. The points of $\operatorname{Sp} R$ are the isomorphism classes of indecomposable injective R-modules, and the collection of subsets $\mathbf{U}_X = \{M \in \operatorname{Sp} R \mid \operatorname{Hom}_R(X, M) = 0\}$, $X \in \operatorname{mod} R$, forms a basis of open subsets for the Zariski topology on $\operatorname{Sp} R$. For example, if R is commutative noetherian, then $\mathbf{X}$ can be identified with the prime spectrum of R and the structure sheaf $\mathcal{O}_{\mathbf{X}}$ is isomorphic to the classical structure sheaf which is defined on the prime spectrum. Moreover, the functor $\operatorname{Mod} R \to \operatorname{Mod} \mathcal{O}_{\mathbf{X}}$, $M \mapsto \widetilde{M}_{\mathbf{X}}$, induces the usual equivalence between $\operatorname{Mod} R$ and the category $\operatorname{Qcoh} \mathbf{X}$ of quasi-coherent sheaves on $(\mathbf{X}, \mathcal{O}_{\mathbf{X}})$; in particular $\Gamma(\mathbf{X}, \widetilde{M}_{\mathbf{X}}) \simeq M$ for every R-module M. However, if R is non-commutative, then the global section functor $M \mapsto \Gamma(\mathbf{X}, \widetilde{M}_{\mathbf{X}})$ has some interesting properties which we discuss in the final part of this chapter.

In these notes we present various aspects of the spectrum of a module category. Some topics are completely independent from each other, but there are also parts which depend on the linear order of our exposition. The material of Chapter 1 and 2 is fundamental; it is needed in each of the subsequent chapters. Chapter 3, 4, 5, 6, 9 are devoted to various topics which are fairly independent from each other. The Chapters 7 and 8 on dimensions form a unit. The discussion of tame algebras in Chapter 10 depends on Chapter 5, 6, 9. The rings of definable scalars from Chapter 11 are needed in Chapter 12 and 13. The Chapters 13, 14, 15 on sheaves form another unit, but Chapter 14 and 15 are independent from each other.

Our treatment of a module category and its spectrum is rather categorical. In fact, we assume that the reader is familiar with the language of abelian Grothendieck categories and its localization theory. An excellent reference is Gabriel's thèse [**26**] or Stenstrøm's book [**79**]. Some extra material is collected in the appendix. Various concepts which arise in this work have their origin in model theory of modules. The standard references are Ziegler's exposition [**80**] and Prest's book [**61**]; an alternative approach can be found in the book of Jensen and Lenzing [**43**]. For the basic facts from representation theory of artin algebras we refer to the textbook of Auslander, Reiten, and Smalø [**8**]. A useful reference for concepts from representation theory of tame algebras is Ringel's book [**71**].

These notes are almost identical with the Habilitationsschrift which I submitted in January 1998 at the University of Bielefeld. There are a number of people I should like to thank for their encouragement and constant interest in my work on the Ziegler spectrum. The first person I should like to mention is Claus Ringel. In fact, I enjoyed very much discussions on the topic of this thesis in the Bielefeld representation theory seminar, and, besides Claus Ringel, various members of this seminar contributed with critics and comments. In particular I should like to thank Peter Dräxler, Lutz Hille, Steffen König, and Jan Schröer. Many thanks also to Helmut Lenzing for stimulating discussions and constant support. My own interest in the functorial approach towards representation theory started when I visited Maurice Auslander at Brandeis in 1991/92; I am grateful to him for his generous contribution of ideas and insight. I should also like to thank Ivo Herzog and Mike Prest for explaining to me the model-theoretic point of view. Parts of this work were done during visits at the universities of Trondheim and Leeds. Many thanks to Idun Reiten and Bill Crawley-Boevey for their hospitality and for stimulating and helpful discussions concerning this material.

CHAPTER 1

The functor category

1.1. Preliminaries

We introduce some terminology. Throughout we are working in a fixed *universe* $\mathfrak{U}$ containing an infinite set (e.g. see [**35**, Numéro 0]). All categories $\mathcal{C}$ are assumed to be $\mathfrak{U}$-*categories* in the sense that for each pair of objects $X, Y \in \mathcal{C}$ the set $\operatorname{Hom}(X, Y)$ is *small*, i.e. bijective with a set in $\mathfrak{U}$. A category $\mathcal{C}$ is *skeletally small* provided that the isomorphism classes of objects in $\mathcal{C}$ form a small set.

Fix a category $\mathcal{A}$. We call the colimit $\varinjlim_{i \in I} X_i$ of a functor $X \colon I \to \mathcal{A}$, $i \mapsto X_i$, a *direct limit* if I is a small directed set. If the category $\mathcal{A}$ has direct limits, then the finitely presented objects of $\mathcal{A}$ play an important role. Recall that an object X in $\mathcal{A}$ is *finitely presented* provided that for every direct limit $\varinjlim Y_i$ in $\mathcal{A}$ the natural morphism $\varinjlim \operatorname{Hom}(X, Y_i) \to \operatorname{Hom}(X, \varinjlim Y_i)$ is an isomorphism. The full subcategory of finitely presented objects of $\mathcal{A}$ is denoted by $\operatorname{fp} \mathcal{A}$. Given a subcategory $\mathcal{C}$ of $\mathcal{A}$ we denote by $\varinjlim \mathcal{C}$ the full subcategory of $\mathcal{A}$ which consists of direct limits $\varinjlim X_i$ with $X_i \in \mathcal{C}$ for all i. The following lemma will be useful.

LEMMA 1.1. *Let $f \colon \mathcal{A} \to \mathcal{B}$ be a functor between categories with direct limits. Suppose there exists a right adjoint which commutes with direct limits. If X is a finitely presented object in $\mathcal{A}$, then $f(X)$ is finitely presented.*

PROOF. Let $\varinjlim Y_i \in \mathcal{B}$. We have the following sequence of isomorphisms

$$\begin{aligned} \varinjlim \operatorname{Hom}(f(X), Y_i) &\simeq \varinjlim \operatorname{Hom}(X, g(Y_i)) \simeq \operatorname{Hom}(X, \varinjlim g(Y_i)) \\ &\simeq \operatorname{Hom}(X, g(\varinjlim Y_i)) \simeq \operatorname{Hom}(f(X), \varinjlim Y_i) \end{aligned}$$

since f and g is a pair of adjoint functors and g commutes with direct limits. The assertion follows. □

Throughout the paper all functors between preadditive categories are assumed to be additive. Given two preadditive categories $\mathcal{C}$ and $\mathcal{D}$, the class of functors $F \colon \mathcal{C} \to \mathcal{D}$ is denoted by $(\mathcal{C}, \mathcal{D})$ and $\operatorname{Hom}(F, G)$ denotes the class of natural transformations between two functors F and G in $(\mathcal{C}, \mathcal{D})$. If $\mathcal{C}$ is skeletally small, then $(\mathcal{C}, \mathcal{D})$ forms actually a category since the Hom sets are small. The category of abelian groups is denoted by Ab. Given any category $\mathcal{C}$, one defines limits, colimits etc. in $(\mathcal{C}, \mathrm{Ab})$ pointwise and they coincide with the categorical notions if $(\mathcal{C}, \mathrm{Ab})$ is a category. If $\mathcal{C}$ is additive, then we denote by

$$\mathcal{C} \longrightarrow (\mathcal{C}, \mathrm{Ab}), \quad X \mapsto H_X = \operatorname{Hom}(X, -)$$

the Yoneda embedding. A functor $F \colon \mathcal{C} \to \mathrm{Ab}$ is said to be *finitely presented* if there is an exact sequence

$$\operatorname{Hom}(Y, -) \longrightarrow \operatorname{Hom}(X, -) \longrightarrow F \longrightarrow 0,$$

equivalently if $F = \operatorname{Coker} H_\phi$ for some map $\phi\colon X \to Y$. We denote by $\operatorname{fp}(\mathcal{C}, \mathrm{Ab})$ the category of finitely presented functors, and observe that for skeletally small $\mathcal{C}$ a functor $F\colon \mathcal{C} \to \mathrm{Ab}$ is finitely presented if and only if $\operatorname{Hom}(F, -)$ preserves direct limits in $(\mathcal{C}, \mathrm{Ab})$.

1.2. Purity

The concept of purity plays an important role in our analysis of a module category. It goes back to Cohn [**13**] and has been further developed by various mathematicians. In this section we collect the basic facts using the embedding of a module category into a bigger functor category.

Let R be an associative ring with identity. Denote by $\operatorname{Mod} R$ the category of (right) R-modules and by $\operatorname{mod} R$ the full subcategory of all finitely presented R-modules. Let $D(R) = (\operatorname{mod} R^{\mathrm{op}}, \mathrm{Ab})$ be the category of covariant additive functors from the category $\operatorname{mod} R^{\mathrm{op}}$ of finitely presented R^{op}-modules into the category Ab of abelian groups. We assign to each R-module M the following three functors:

$$\begin{aligned} H_M &\colon \operatorname{Mod} R \longrightarrow \mathrm{Ab}, && X \mapsto \operatorname{Hom}_R(M, X) \\ T_M &\colon \operatorname{Mod} R^{\mathrm{op}} \longrightarrow \mathrm{Ab}, && X \mapsto M \otimes_R X \\ E_M &\colon D(R) \longrightarrow \mathrm{Ab}, && X \mapsto \operatorname{Hom}_{D(R)}(X, T_M) \end{aligned}$$

Of course, these assignments are functorial, i.e. each map $M \to N$ between R-modules gives rise to natural transformations $H_N \to H_M$, $T_M \to T_N$, and $E_M \to E_N$. We shall often consider restrictions of these functors. However, our notation will not distinguish between the original functor and its restriction if the context is clear.

The fully faithful functor

$$\operatorname{Mod} R \longrightarrow D(R), \quad M \mapsto T_M = M \otimes_R -$$

will play an important role in our considerations. We recall briefly the basic facts about $D(R)$ and refer the reader for a detailed exposition to [**38**]. A sequence (finite or infinite) of R-module maps

$$\cdots \longrightarrow M_{n-1} \longrightarrow M_n \longrightarrow M_{n+1} \longrightarrow \cdots$$

is said to be *pure-exact* if its image under the functor $M \mapsto T_M$

$$\cdots \longrightarrow T_{M_{n-1}} \longrightarrow T_{M_n} \longrightarrow T_{M_{n+1}} \longrightarrow \cdots$$

is exact in $D(R)$. In particular, a map $M \to N$ is a *pure monomorphism* if $T_M \to T_N$ is a monomorphism. An R-module M is called *pure-injective* if every pure monomorphism $M \to N$ splits. We shall also use the full subcategory $C(R) = \operatorname{fp}(\operatorname{mod} R^{\mathrm{op}}, \mathrm{Ab})$ of finitely presented functors in $D(R)$. Recall that a functor in $D(R)$ is *finitely presented* if there is an exact sequence $\operatorname{Hom}_{R^{\mathrm{op}}}(Y, -) \to \operatorname{Hom}_{R^{\mathrm{op}}}(X, -) \to F \to 0$ with X and Y in $\operatorname{mod} R^{\mathrm{op}}$. Observe that $C(R)$ is an exact abelian subcategory, equivalently that finitely generated subfunctors of finitely presented functors again are finitely presented. This follows from the fact that $\operatorname{mod} R^{\mathrm{op}}$ is a category having cokernels. Viewing the ring R as a category with one object, the isomorphism $R \simeq \operatorname{End}(T_R)$ amounts to a fully faithful functor $i\colon R \to C(R)$. The next lemma expresses the fact that $C(R)$ is the free abelian category over R, see [**36**].

LEMMA 1.2. *Any additive functor* $f\colon R \to \mathcal{A}$ *into an abelian category* $\mathcal{A}$ *extends, up to isomorphism, uniquely to an exact functor* $f'\colon C(R) \to \mathcal{A}$ *with* $f = f' \circ i$.

PROOF. First one extends f to a contravariant additive functor $f_1\colon \operatorname{proj} R^{\mathrm{op}} \to \mathcal{A}$ on the category of finitely generated projective R^{op}-modules. This extends to a left exact functor $f_2\colon \operatorname{mod} R^{\mathrm{op}} \to \mathcal{A}$ by $f_2(X) = \operatorname{Ker} f_1(\phi)$ for $X = \operatorname{Coker}\phi$ with ϕ a map in $\operatorname{proj} R^{\mathrm{op}}$. Finally, one extends f_2 to an exact functor $f'\colon C(R) \to \mathcal{A}$ by $f'(F) = \operatorname{Coker} f_2(\psi)$ for $F = \operatorname{Coker}\operatorname{Hom}_{R^{\mathrm{op}}}(\psi, -)$ with ψ a map in $\operatorname{mod} R^{\mathrm{op}}$. □

Recall that an object M in any abelian category with direct limits is *fp-injective* if $\operatorname{Ext}^1(X, M) = 0$ for every finitely presented object X. We shall use the following characterization of fp-injective objects in $D(R)$.

LEMMA 1.3. *The following are equivalent for a functor* F *in* $D(R)$*:*

(1) F *is fp-injective, i.e.* $\operatorname{Ext}^1(-, F)$ *vanishes on* $C(R)$.
(2) $\operatorname{Hom}(-, F)$ *is an exact functor on* $C(R)$.
(3) $F \simeq T_M$ *for some* R*-module* M.

PROOF. (1) ⇒ (2) An exact sequence $0 \to G' \to G \to G'' \to 0$ in $C(R)$ induces an exact sequence $0 \to \operatorname{Hom}(G'', F) \to \operatorname{Hom}(G, F) \to \operatorname{Hom}(G', F) \to \operatorname{Ext}^1(G'', F)$. Therefore $\operatorname{Hom}(-, F)$ is exact provided that $\operatorname{Ext}^1(-, F) = 0$.

(2) ⇒ (3) Suppose that $\operatorname{Hom}(-, F)$ is exact of $C(R)$ and let $M = F(R)$ with R acting on M via $R \simeq \operatorname{End}_{R^{\mathrm{op}}}(R)$. The isomorphism $T_M(R) \simeq F(R)$ extends to an isomorphism $T_M(X) \simeq F(X)$ for every X in $\operatorname{mod} R^{\mathrm{op}}$. This follows from Yoneda's lemma and the fact that a presentation $R^n \to R^m \to X \to 0$ induces an exact sequence $0 \to \operatorname{Hom}_{R^{\mathrm{op}}}(X, -) \to \operatorname{Hom}_{R^{\mathrm{op}}}(R^m, -) \to \operatorname{Hom}_{R^{\mathrm{op}}}(R^n, -)$ in $C(R)$. Thus $T_M \simeq F$.

(3) ⇒ (1) Let $F \simeq T_M$ and $G \in C(R)$. There exists a projective presentation $0 \to \operatorname{Hom}_{R^{\mathrm{op}}}(Z, -) \to \operatorname{Hom}_{R^{\mathrm{op}}}(Y, -) \to \operatorname{Hom}_{R^{\mathrm{op}}}(X, -) \to G \to 0$ which is induced by an exact sequence $X \to Y \to Z \to 0$ in $\operatorname{mod} R^{\mathrm{op}}$. The sequence remains exact after applying $\operatorname{Hom}(-, F)$ since T_M is right exact. Therefore $\operatorname{Ext}^1(G, F) = 0$. □

Every R-module can be viewed as a functor $R^{\mathrm{op}} \to \mathrm{Ab}$, i.e. $\operatorname{Mod} R = (R^{\mathrm{op}}, \mathrm{Ab})$. Combining this fact with the universal property of $C(R)$ stated in Lemma 1.2 one obtains the following reformulation of the preceding lemma.

LEMMA 1.4. *The functor* $\operatorname{Mod} R \to \operatorname{Ex}(C(R)^{\mathrm{op}}, \mathrm{Ab})$, $M \mapsto E_M$, *is an equivalence.*

Here we denote for any abelian category $\mathcal{C}$ by $\operatorname{Ex}(\mathcal{C}, \mathrm{Ab})$ the category of exact functors $\mathcal{C} \to \mathrm{Ab}$. Note that $F \mapsto F(T_R)$ is an inverse for $M \mapsto E_M$ where R acts on $F(T_R)$ via the isomorphism $R \simeq \operatorname{End}(T_R)$.

Having characterized the fp-injective objects in $D(R)$ we can now prove that the assignment $M \mapsto T_M$ identifies the pure-injective R-modules with the injective objects in $D(R)$.

LEMMA 1.5. *The following are equivalent for a functor* F *in* $D(R)$*:*

(1) F *is injective, i.e.* $\operatorname{Ext}^1(-, F)$ *vanishes on* $D(R)$.
(2) $\operatorname{Hom}(-, F)$ *is an exact functor on* $D(R)$.
(3) $F \simeq T_M$ *for some pure-injective* R*-module* M.

PROOF. (1) $\Leftrightarrow$ (2) This is standard.

(1) $\Rightarrow$ (3) Any injective object is fp-injective and therefore $F \simeq T_M$ for some R-module M by Lemma 1.3. The module M is pure-injective since every monomorphism $F \to T_N$ in $D(R)$ splits.

(3) $\Rightarrow$ (1) Choose a monomorphism $F \to G$ with G injective in $D(R)$. This maps splits since $G \simeq T_N$ for some R-module N and M is pure-injective. Therefore F is injective. $\square$

We discuss now a duality between $C(R)$ and $C(R^{\text{op}})$ which extends the duality between R and R^{op}. Given a functor $F \in D(R)$ we define $F^\vee \colon \operatorname{Mod} R \to \text{Ab}$ by $F^\vee(X) = \operatorname{Hom}_{D(R)}(F, T_X)$. The same formula defines for every $F \in D(R^{\text{op}})$ a functor $F^\vee \colon \operatorname{Mod} R^{\text{op}} \to \text{Ab}$.

LEMMA 1.6. *The following holds:*

(1) *The assignment $F \mapsto F^\vee$ induces mutually inverse equivalences between $C(R)^{\text{op}}$ and $C(R^{\text{op}})$.*
(2) $(\operatorname{Coker} H_\phi)^\vee = \operatorname{Ker} T_\phi$ *for every map ϕ in* $\operatorname{mod} R$.
(3) *$F \in D(R)$ is finitely presented if and only if $F = \operatorname{Ker} T_\phi$ for some map ϕ in* $\operatorname{mod} R$.

PROOF. (1) The functor $C(R) \to D(R^{\text{op}})$, $F \mapsto F^\vee$, sends T_R to $H_R = T_{R^{\text{op}}}$ and is exact by Lemma 1.3. It is easily checked that $F^\vee \in C(R^{\text{op}})$ for every $F \in C(R)$. For instance, $(H_X)^\vee = T_X$ for every X in $\operatorname{mod} R^{\text{op}}$ and a presentation $R^n \xrightarrow{\phi} R^m \to X \to 0$ shows that $T_X = \operatorname{Ker} T_\phi$ is finitely presented since it is the kernel of a map between finitely presented functors. It follows from Lemma 1.2 that $C(R) \to C(R)$, $F \mapsto F^{\vee\vee}$, is isomorphic to the identity functor. Thus $^\vee$ induces a duality between $C(R)$ and $C(R^{\text{op}})$.

(2) $(H_X)^\vee = T_X$ for every X in $\operatorname{mod} R^{\text{op}}$, and $(\operatorname{Coker} H_\phi)^\vee = \operatorname{Ker} T_\phi$ follows from the exactness of $^\vee$.

(3) follows from (1) - (2). $\square$

A functor $F \colon \operatorname{Mod} R \to \text{Ab}$ defined on the category of all R-modules will be called *coherent* if $F = \operatorname{Coker} H_\phi$ for some map ϕ between finitely presented R-modules.

CHAPTER 2

Definable subcategories

This chapter is devoted to the fundamental concept of a definable subcategory of $\operatorname{Mod} R$. This notion has its origin in model theory of modules; in this context a definable subcategory corresponds to a complete theory of modules [**61, 80**]. Definable subcategories were introduced formally by Crawley-Boevey when their relevance became apparent in representation theory [**19**].

2.1. Definable subcategories

Let $\phi\colon X \to Y$ be a map of R-modules. An R-module M is called *ϕ-injective* if the induced map $\operatorname{Hom}_R(Y, M) \to \operatorname{Hom}_R(X, M)$ is surjective. If Φ is a collection of maps, then M is Φ-injective if M is ϕ-injective for every ϕ in Φ, and we denote by $(\operatorname{Mod} R)_\Phi$ the full subcategory of Φ-injective R-modules. A full subcategory of $\operatorname{Mod} R$ formed by the Φ-injective modules for some collection Φ of maps in $\operatorname{mod} R$ is called *definable*.

We now formulate the main result of this section. It is due to Crawley-Boevey [**19**], with the exception of part (4) which is crucial for the proof given here. Note that Crawley-Boevey's proof uses model-theoretic arguments, including some work of Herzog [**40**] and Ziegler [**80**].

THEOREM 2.1. *For a full subcategory $\mathcal{X}$ of* $\operatorname{Mod} R$ *the following are equivalent.*

(1) *$\mathcal{X}$ is definable.*
(2) *$\mathcal{X}$ is closed under taking direct limits, products and pure submodules in* $\operatorname{Mod} R$.
(3) *There is a Serre subcategory $\mathcal{S}$ of $C(R)$ such that an R-module M belongs to $\mathcal{X}$ if and only if* $\operatorname{Hom}(\mathcal{S}, T_M) = 0$.
(4) *There is a hereditary torsion theory $(\mathcal{T}, \mathcal{F})$ of finite type for $D(R)$ such that an R-module M belongs to $\mathcal{X}$ if and only if T_M is in $\mathcal{F}$.*

Recall that a full subcategory $\mathcal{S}$ of $C(R)$ is a *Serre subcategory* provided that for every exact sequence $0 \to F' \to F \to F'' \to 0$ in $C(R)$ the object F belongs to $\mathcal{S}$ if and only if F' and F'' belong to $\mathcal{S}$. A torsion theory $(\mathcal{T}, \mathcal{F})$ for $D(R)$ is *hereditary* if $\mathcal{T}$ is closed under subobjects and $(\mathcal{T}, \mathcal{F})$ is said to be of *finite type* if $\mathcal{F}$ is closed under direct limits. We shall use the fact that $(\mathcal{T}, \mathcal{F})$ is hereditary of finite type if and only if it is of the form $(\varinjlim \mathcal{S}, \mathcal{F})$ for some Serre subcategory $\mathcal{S}$ of $C(R)$, see Proposition A.4. Here, $\varinjlim \mathcal{S}$ denotes the full subcategory of objects in $D(R)$ which consists of direct limits $\varinjlim F_i$ with F_i in $\mathcal{S}$ for all i.

The next lemma is a useful reformulation of the definition of ϕ-injectivity.

LEMMA 2.2. *The following are equivalent for a map $\phi \in \operatorname{mod} R$ and $M \in$* $\operatorname{Mod} R$*:*

(1) *M is ϕ-injective.*

(2) $\operatorname{Hom}(\operatorname{Ker} T_\phi, T_M) = 0$.
(3) $\operatorname{Coker} H_\phi(M) = 0$.

PROOF. Clear. □

PROOF OF THEOREM 2.1. (1) $\Leftrightarrow$ (3) We use the preceding lemma. Suppose first that $\mathcal{X}$ is the category of Φ-injective modules for some collection Φ of maps in $\operatorname{mod} R$. Denote by $\mathcal{S}$ the Serre subcategory of $C(R)$ which is generated by the collection of functors $\operatorname{Ker} T_\phi$ with $\phi \in \Phi$. It follows from the fp-injectivity of each T_M that $M \in \mathcal{X}$ if and only if $\operatorname{Hom}(\mathcal{S}, T_M) = 0$. Conversely, suppose that (3) holds. Let Φ be the collection of maps in $\operatorname{mod} R$ with $\operatorname{Ker} T_\phi \in \mathcal{S}$. Using the fact that every functor in $C(R)$ is of the form $\operatorname{Ker} T_\phi$ for some map ϕ in $\operatorname{mod} R$, it follows that $M \in \mathcal{X}$ if and only if M is Φ-injective. Therefore $\mathcal{X}$ is definable.

(1) $\Rightarrow$ (2) We use again the preceding lemma. It is well-known that H_X preserves direct limits and products whenever X is a finitely presented module. Therefore $\operatorname{Coker} H_\phi$ preserves direct limits and products for any map ϕ in $\operatorname{mod} R$. It follows that $\mathcal{X}$ is closed under taking direct limits and products. To see that $\mathcal{X}$ is closed under taking pure submodules one uses part (2) of the preceding lemma.

(2) $\Rightarrow$ (4) Consider the class $\mathcal{F} = \{F \in D(R) \mid F \subseteq T_M \text{ for some } M \in \mathcal{X}\}$. We claim that $\mathcal{F}$ is the torsion-free class for some hereditary torsion theory of finite type for $D(R)$. We want to show that $\mathcal{F}$ is closed under taking direct limits, products and subobjects. It is clear that $\mathcal{F}$ is closed under taking products and subobjects. Therefore suppose that $F = \varinjlim_{i \in I} F_i$ is a direct limit of objects F_i in $\mathcal{F}$ and fix for each i a monomorphism $F_i \to G_i$ with $G_i = T_{M_i}$ for some $M_i \in \mathcal{X}$. For each $i \in I$ let $J_i = \{j \in I \mid i \leq j\}$. The canonically defined monomorphisms $F_i \to \prod_{j \in J_i} G_j$, $i \in I$, induce a monomorphism $\varinjlim_{i \in I} F_i \to \varinjlim_{i \in I} \prod_{j \in J_i} G_j$ since direct limits in $D(R)$ are left exact. Defining $M = \varinjlim_{i \in I} \prod_{j \in J_i} M_j$ we have $T_M = \varinjlim_{i \in I} \prod_{j \in J_i} G_j$ since the functor $N \mapsto T_N$ preserves direct limits and products. We obtain therefore a monomorphism $F \to T_M$ with M in $\mathcal{X}$, and this shows that $\mathcal{F}$ is also closed under direct limits. Now consider $\mathcal{F}$ as a full subcategory of $D(R)$. The inclusion functor $\mathcal{F} \to D(R)$ has a left adjoint $f\colon D(R) \to \mathcal{F}$ which is constructed as follows. For $F \in D(R)$ let G_i, $i \in I$, be the set of quotient objects of F which are in $\mathcal{F}$. Define $f(F)$ to be the image and $t(F)$ the kernel of the canonical morphism $F \to \prod_i G_i$. Also define

$$\begin{aligned}\mathcal{T} &= \{F \in D(R) \mid \operatorname{Hom}(F, \mathcal{F}) = 0\} \\ &= \{F \in D(R) \mid \operatorname{Hom}(F, T_M) = 0 \text{ for all } M \in \mathcal{X}\}\end{aligned}$$

and note that $\mathcal{S} = \mathcal{T} \cap C(R)$ is a Serre subcategory of $C(R)$. We claim that $t(F) \in \varinjlim \mathcal{S}$ for $F \in C(R)$. To this end write $t(F) = \varinjlim G_i$ as direct limit of finitely generated subobjects. We need to show that $G_i \in \mathcal{S}$ for all i. Suppose that $G = G_i \notin \mathcal{S}$. Then there is a non-zero morphism $\phi\colon G \to T_M$ for some $M \in \mathcal{X}$ and ϕ extends to a morphism $\psi\colon F \to T_M$ since $\operatorname{Ext}^1(F/G, T_M) = 0$. But the adjointness property of f implies that ψ factors through $F \to f(F)$. Therefore $\phi(G) = 0$, a contradiction to our assumption. Thus $t(F) \in \varinjlim \mathcal{S}$. Now let $F = \varinjlim F_i$ be an arbitrary object in $D(R)$ written as direct limit of finitely presented objects. We obtain an exact sequence $0 \to \varinjlim t(F_i) \to F \to \varinjlim f(F_i) \to 0$ with $\varinjlim t(F_i) \in \varinjlim \mathcal{S}$ and $\varinjlim f(F_i) \in \mathcal{F}$ since both $\varinjlim \mathcal{S}$ and $\mathcal{F}$ are closed under direct limits. We conclude that $(\varinjlim \mathcal{S}, \mathcal{F})$ is a torsion theory for $D(R)$ which is hereditary of finite type.

(4) $\Rightarrow$ (3) Suppose we have $\mathcal{X} = \{M \in \operatorname{Mod} R \mid T_M \in \mathcal{F}\}$ for some hereditary torsion theory $(\mathcal{T}, \mathcal{F})$ of finite type for $D(R)$. The assertion follows immediately if we take $\mathcal{S} = \mathcal{T} \cap C(R)$ as defining subcategory for $\mathcal{X}$ since $\mathcal{F} = \{F \in D(R) \mid \operatorname{Hom}(\mathcal{S}, F) = 0\}$. □

Given a definable subcategory $\mathcal{X} = (\operatorname{Mod} R)_\Phi$ of $\operatorname{Mod} R$, we denote by $\mathcal{T}_\mathcal{X}$ or $\mathcal{T}_\Phi$ the localizing subcategory of functors F in $D(R)$ such that $\operatorname{Hom}(F, T_M) = 0$ for all M in $\mathcal{X}$. The Serre subcategory $\mathcal{T}_\mathcal{X} \cap C(R)$ is denoted by $\mathcal{S}_\mathcal{X}$ or $\mathcal{S}_\Phi$. The following result is a consequence of the preceding theorem and its proof.

COROLLARY 2.3. *The assignments*

$$\mathcal{X} \mapsto \mathcal{S}_\mathcal{X} = \mathcal{T}_\mathcal{X} \cap C(R) \quad \text{and} \quad \mathcal{S} \mapsto \{M \in \operatorname{Mod} R \mid \operatorname{Hom}(\mathcal{S}, T_M) = 0\}$$

give mutually inverse bijections between the definable subcategories of $\operatorname{Mod} R$ *and the Serre subcategories of* $C(R)$.

The characterization of definable subcategories in Theorem 2.1 leads to further characterizations. In fact, using the basic theory of localization in the locally coherent category $D(R)$ which is formulated in the appendix, we obtain the following consequence of Theorem 2.1.

COROLLARY 2.4. *For a full subcategory* $\mathcal{X}$ *of* $\operatorname{Mod} R$ *the following are equivalent:*

(1) *$\mathcal{X}$ is a definable subcategory.*
(2) *There is a Serre subcategory $\mathcal{S}$ of $C(R)$ such that the assignment $M \mapsto E_M$ induces an equivalence between $\mathcal{X}$ and the category $\operatorname{Ex}((C(R)/\mathcal{S})^{\mathrm{op}}, \mathrm{Ab})$ of exact functors $(C(R)/\mathcal{S})^{\mathrm{op}} \to \mathrm{Ab}$.*
(3) *There is a localizing subcategory of finite type $\mathcal{T}$ of $D(R)$ such that the composition of the section functor $D(R)/\mathcal{T} \to D(R)$ with $D(R) \to \operatorname{Mod} R$, $F \mapsto F(R)$, induces an equivalence between the full subcategory of fp-injective objects in $D(R)/\mathcal{T}$ and the category $\mathcal{X}$.*

REMARK 2.5. If the conditions (1) - (3) in the preceding theorem are satisfied, then $\mathcal{S} = \mathcal{S}_\mathcal{X}$ and $\mathcal{T} = \mathcal{T}_\mathcal{X}$.

PROOF. (1) $\Leftrightarrow$ (2) We combine the characterization of a definable subcategory in part (3) of Theorem 2.1 with the equivalence $\operatorname{Mod} R \to \operatorname{Ex}(C(R)^{\mathrm{op}}, \mathrm{Ab})$ from Lemma 1.4. The assertion follows from the fact that for any Serre subcategory $\mathcal{S}$ of $C(R)$ with quotient functor $p\colon C(R) \to C(R)/\mathcal{S}$, there exists a functor F in $\operatorname{Ex}((C(R)/\mathcal{S})^{\mathrm{op}}, \mathrm{Ab})$ with $E_M = F \circ p$ for any R-module M if and only if $\operatorname{Hom}(\mathcal{S}, T_M) = 0$.

(1) $\Leftrightarrow$ (3) Combine the characterization of a definable subcategory in part (4) of Theorem 2.1 with Proposition A.7. □

We have seen in Lemma 1.5 that the functor $\operatorname{Mod} R \to D(R)$, $M \mapsto T_M$, induces an equivalence between $\operatorname{Mod} R$ and the full subcategory of fp-injective objects in $D(R)$. This is a special case of the following result which is fundamental for our analysis of definable subcategories of $\operatorname{Mod} R$.

COROLLARY 2.6. *Let $\mathcal{X}$ be a definable subcategory of* $\operatorname{Mod} R$. *Then the functor*

$$\operatorname{Mod} R \longrightarrow D(R) \longrightarrow D(R)/\mathcal{T}_\mathcal{X}, \quad M \mapsto T_M$$

induces equivalences:

(1) *between $\mathcal{X}$ and the full subcategory of fp-injective objects in $D(R)/\mathcal{T}_{\mathcal{X}}$;*
(2) *between the full subcategory of pure-injective modules in $\mathcal{X}$ and the full subcategory of injective objects in $D(R)/\mathcal{T}_{\mathcal{X}}$.*

PROOF. The quotient functor $q\colon D(R) \to D(R)/\mathcal{T}_{\mathcal{X}}$ is an inverse for the section functor $s\colon D(R)/\mathcal{T}_{\mathcal{X}} \to D(R)$, i.e. $q \circ s \simeq \mathrm{id}$. Therefore the assertion of the corollary follows from the description of $\mathcal{X}$ in part (4) of Theorem 2.1, in combination with Proposition A.7 and Proposition A.8. □

We give an application of the preceding result. To this end recall that a module M is Σ-pure-injective if every coproduct of copies of M is pure-injective.

COROLLARY 2.7. *The following are equivalent for a definable subcategory $\mathcal{X}$ of* $\operatorname{Mod} R$*:*

(1) *Every module in $\mathcal{X}$ is pure-injective.*
(2) *Every module in $\mathcal{X}$ is Σ-pure-injective.*
(3) *Every module in $\mathcal{X}$ is a coproduct of indecomposable modules.*

PROOF. Using the description of a definable subcategory from the preceding corollary, it follows from Proposition A.11 that each condition is equivalent to $D(R)/\mathcal{T}$ being locally noetherian. □

2.2. Saturation

Given a collection Φ of maps in $\operatorname{mod} R$, we now describe the maximal collection Ψ such that $(\operatorname{Mod} R)_\Phi = (\operatorname{Mod} R)_\Psi$. We begin with a definition. Let $\mathcal{C}$ be an additive category with cokernels, so that $\operatorname{fp}(\mathcal{C}, \mathrm{Ab})$ is abelian. A collection Φ of maps in $\mathcal{C}$ will be called *Serre saturated*, or simply *saturated*, if the following conditions hold:

(S1) Suppose there are maps $\alpha_i, \beta_i, \phi_i$ $(i = 1, 2)$ in $\mathcal{C}$ satisfying the identities $\phi_1 \circ \alpha_1 = \beta_2 \circ \phi_2$ and $\phi_2 \circ \alpha_2 = \beta_1 \circ \phi_1$. Suppose also that $\mathrm{id} - \alpha_2 \circ \alpha_1$ factors through ϕ_2. Then $\phi_1 \in \Phi$ implies $\phi_2 \in \Phi$.
(S2) Suppose there are maps $\phi_i\colon X_i \to Y_i$ $(i = 1, 2)$ and $\psi\colon X_1 \to Y_2$ such that the composition $X_1 \xrightarrow{\psi} Y_2 \to \operatorname{Coker} \phi_2$ is zero. Then $\phi_1, \phi_2 \in \Phi$ if and only if $\left[\begin{smallmatrix} \phi_1 & 0 \\ \psi & \phi_2 \end{smallmatrix}\right] \in \Phi$.

The following lemma explains our definition of a saturated collection of maps.

LEMMA 2.8. *For a collection Φ of maps in $\mathcal{C}$ the following are equivalent:*

(1) *Φ is Serre saturated.*
(2) *$\{\operatorname{Coker} H_\phi \mid \phi \in \Phi\}$ is a Serre subcategory of* $\operatorname{fp}(\mathcal{C}, Ab)$.

PROOF. We fix two maps $\phi_i\colon X_i \to Y_i$ $(i = 1, 2)$. (S1) expresses the fact that $\phi_2 \in \Phi$ provided that $\phi_1 \in \Phi$ and $\operatorname{Coker} H_{\phi_2}$ is a direct summand of $\operatorname{Coker} H_{\phi_1}$. There exists an exact sequence $0 \to \operatorname{Coker} H_{\phi_1} \to F \to \operatorname{Coker} H_{\phi_2} \to 0$ if and only if there is a map $\psi\colon X_1 \to Y_2$ such that the composition $X_1 \xrightarrow{\psi} Y_2 \to \operatorname{Coker} \phi_2$ is zero and $\operatorname{Coker} H_\phi \simeq F$ for $\phi = \left[\begin{smallmatrix} \phi_1 & 0 \\ \psi & \phi_2 \end{smallmatrix}\right]$, see Lemma C.5. Therefore, assuming (S1), condition (S2) holds if and only if (2) holds. □

Given any collection Φ of maps in $\mathcal{C}$, we denote by $\overline{\Phi}$ the smallest saturated collection of maps containing Φ and call $\overline{\Phi}$ the *saturation* of Φ. This definition makes sense because any intersection $\bigcap \Phi_i$ of saturated collections Φ_i is again saturated. Using the preceding lemma we obtain the following consequence of Corollary 2.3.

THEOREM 2.9. *Let Φ be a collection of maps in* $\operatorname{mod} R$. *Then a map ϕ in* $\operatorname{mod} R$ *belongs to the saturation $\overline{\Phi}$ if and only if M is ϕ-injective for every Φ-injective R-module M. Therefore the assignment $\Psi \mapsto (\operatorname{Mod} R)_\Psi$ gives a bijection between the saturated collections of maps in* $\operatorname{mod} R$ *and the definable subcategories of* $\operatorname{Mod} R$.

PROOF. A map ϕ in $\operatorname{mod} R$ belongs to $\overline{\Phi}$ if and only if $\operatorname{Ker} T_\phi \in \mathcal{S}_\Phi$. Therefore the bijection between saturated collections of maps and definable subcategories is a reformulation of Corollary 2.3. □

Given any class $\mathcal{C}$ of R-modules we denote by $\Phi_\mathcal{C}$ the collection of maps $\phi\colon X \to Y$ in $\operatorname{mod} R$ such that every map $X \to M$ with M in $\mathcal{C}$ factors through ϕ. Note that $\Phi_\mathcal{C}$ is automatically saturated. Also, $(\operatorname{Mod} R)_{\Phi_\mathcal{C}}$ is the smallest definable subcategory containing $\mathcal{C}$; we call it therefore the definable subcategory *generated by* $\mathcal{C}$.

2.3. The Ziegler topology

The isomorphism classes of indecomposable pure-injective R-modules form a set which we denote by $\operatorname{Ind} R$. In fact, the functor T_M corresponding to an indecomposable pure-injective R-module M is the injective envelope of a quotient T_R/U if one takes $U = \operatorname{Ker} T_\phi$ for some non-zero map $\phi\colon R \to M$. Therefore

$$\operatorname{card} \operatorname{Ind} R \leq 2^{\aleph_0 + \operatorname{card} R}$$

since every subfunctor of T_R is determined by a set of maps in $\operatorname{mod} R$. In [**80**] Ziegler introduced a topology on the set $\operatorname{Ind} R$ which can be described as follows.

PROPOSITION 2.10. *The subsets of the form*

$$\mathbf{U}_\Phi = \{M \in \operatorname{Ind} R \mid M \textit{ is } \Phi\textit{-injective}\}$$

(Φ a collection of maps in $\operatorname{mod} R$*) form the closed sets of a topology on* $\operatorname{Ind} R$. *An open subset* $\operatorname{Ind} R \setminus \mathbf{U}_\Phi$ *is quasi compact if and only if* $\mathbf{U}_\Phi = \mathbf{U}_\phi$ *for some map ϕ in* $\operatorname{mod} R$*; in particular* $\operatorname{Ind} R$ *is quasi-compact.*

We call a subset $\mathbf{U}$ of $\operatorname{Ind} R$ *Ziegler-closed* if $\mathbf{U} = \mathbf{U}_\Phi$ for some Φ in $\operatorname{mod} R$, and the complement of a Ziegler-closed subset is called *Ziegler-open.* The set $\operatorname{Ind} R$, equipped with the Ziegler topology, is called the *Ziegler spectrum* of R.

PROOF OF PROPOSITION 2.10. We have $\operatorname{Ind} R = \mathbf{U}_\emptyset$ and $\emptyset = \mathbf{U}_\phi$ for $\phi\colon R \to 0$. Now let $(\Phi_i)_{i\in I}$ be a family of collections of maps in $\operatorname{mod} R$, and denote for each i by $\mathcal{S}_i = \mathcal{S}_{\Phi_i}$ the Serre subcategory of $C(R)$ corresponding to Φ_i. Clearly, $\bigcap \mathbf{U}_{\Phi_i} = \mathbf{U}_\Phi$ for $\Phi = \bigcup \Phi_i$. If I is finite, let $\Psi = \{\phi \in \operatorname{mod} R \mid \operatorname{Ker} T_\phi \in \mathcal{S}\}$ where $\mathcal{S} = \bigcap \mathcal{S}_i$. It is not hard to check that $\bigcup \mathbf{U}_{\Phi_i} = \mathbf{U}_\Psi$, see [**50**, Lemma 4.1].

It remains to verify the characterization of the quasi-compact open subsets $\mathbf{V} = \operatorname{Ind} R \setminus \mathbf{U}_\Phi$. Suppose first that $\mathbf{U}_\Phi = \mathbf{U}_\phi$ for some ϕ, and let $\mathbf{V} = \bigcup_{i\in I} \mathbf{V}_i$ be an open covering with $\mathbf{V}_i = \operatorname{Ind} R \setminus \mathbf{U}_{\Phi_i}$ for all i. We have $\mathcal{S}_\phi = \bigcup_{i\in I} \mathcal{S}_i$ and this implies $F = \operatorname{Ker} T_\phi \in \bigcup_{i\in J} \mathcal{S}_i$ for some finite $J \subseteq I$ since $F \in \bigcup_{i\in I} \mathcal{S}_i$ if and only if there exists a finite chain

$$0 = F_0 \subseteq F_1 \subseteq \ldots \subseteq F_n = F$$

in $C(R)$ with elements $i_j \in I$ such that $F_{j+1}/F_j \in \mathcal{S}_{i_j}$ for all j. Thus $\mathcal{S}_\phi = \bigcup_{i\in J} \mathcal{S}_i$ since $\mathcal{S}_\phi$ is generated by F, and we conclude that $\mathbf{V} = \bigcup_{i\in J} \mathbf{V}_i$. It follows that $\mathbf{V}$ is quasi-compact. To prove the converse consider the open covering $\mathbf{V} = \bigcup_{\psi\in\Phi} \mathbf{V}_\psi$

with $\mathbf{V}_\psi = \operatorname{Ind} R \setminus \mathbf{U}_\psi$. If there is a finite subset $\Psi \subseteq \Phi$ with $\mathbf{V} = \bigcup_{\psi\in\Psi} \mathbf{V}_\psi$, then $\mathbf{U}_\Phi = \mathbf{U}_\phi$ for $\phi = \amalg_{\psi\in\Psi}\psi$. □

We present now a remarkable property of the Ziegler-closed subset $\mathbf{U}_\Phi$ of $\operatorname{Ind} R$ which corresponds to a definable subcategory $(\operatorname{Mod} R)_\Phi$.

THEOREM 2.11. *An R-module M is Φ-injective if and only if M is isomorphic to a pure submodule of some product $\prod_{i\in I} M_i$ with $M_i \in \mathbf{U}_\Phi$ for all i.*

PROOF. Let $\mathcal{S}_\Phi$ be the Serre subcategory corresponding to Φ. Then M is Φ-injective if and only if $\operatorname{Hom}(\mathcal{S}_\Phi, T_M) = 0$, and therefore the assertion is a consequence of Proposition A.9. □

We obtain as a direct consequence a result which is due to Crawley-Boevey [**19**] and implicit in Ziegler's work [**80**].

COROLLARY 2.12. *Two definable subcategories of* $\operatorname{Mod} R$ *coincide if and only if they contain the same indecomposable pure-injective modules.*

Another consequence of Theorem 2.11 is obtained from the bijective correspondence between definable subcategories of $\operatorname{Mod} R$ and Serre subcategories of $C(R)$. This result is due to Herzog [**40**], and the first proof using functor categories appears in [**50**].

COROLLARY 2.13. *The assignments*

$$\mathbf{U} \mapsto \{F \in C(R) \mid \operatorname{Hom}(F, T_M) = 0 \text{ for every } M \in \mathbf{U}\} \quad \text{and}$$

$$\mathcal{S} \mapsto \{M \in \operatorname{Ind} R \mid \operatorname{Hom}(\mathcal{S}, T_M) = 0\}$$

give mutually inverse bijections between the Ziegler-closed subsets of $\operatorname{Ind} R$ *and the Serre subcategories of* $C(R)$.

There are various classes of rings where a complete description of the Ziegler spectrum is known; we refer the reader to [**80, 61, 66, 73**].

2.4. Definable quotient categories

A discussion of definable subcategories would not be complete without at least mentioning the corresponding concept of a definable quotient category which was introduced in [**52**], see also [**49**]. In this section we recall briefly the definition and some of the essential properties. Fix a definable subcategory $\mathcal{X}$ of $\operatorname{Mod} R$ with $\mathcal{S} = \mathcal{S}_\mathcal{X}$ and $\mathcal{T} = \mathcal{T}_\mathcal{X}$. It was shown in Corollary 2.4 that $\mathcal{X}$ and $\operatorname{Ex}(C(\mathcal{X})^{\mathrm{op}}, \mathrm{Ab})$ are equivalent via the functor $M \mapsto \operatorname{Hom}(-, T_M)$ where $C(\mathcal{X}) = C(R)/\mathcal{S}$. Therefore we call $\mathcal{Z} = \operatorname{Ex}(\mathcal{S}^{\mathrm{op}}, \mathrm{Ab})$ the *definable quotient category* of $\operatorname{Mod} R$ with respect to $\mathcal{X}$, and $q\colon \operatorname{Mod} R \to \mathcal{Z}$, $M \mapsto \operatorname{Hom}(-, T_M)|_\mathcal{S}$, is the corresponding *quotient functor*. Note that q is an equivalence if and only if $\mathcal{X} = 0$; we use therefore the notation $C(\mathcal{Z}) = \mathcal{S}$. It is clear that $\mathcal{Z}$ is an additive category with direct limits and products. There is also a notion of purity for $\mathcal{Z}$ which specializes to the usual one if $\mathcal{X} = 0$. A sequence of morphisms $0 \to L \to M \to N \to 0$ in $\mathcal{Z}$ is *pure-exact* if $0 \to L(X) \to M(X) \to N(X) \to 0$ is exact for all X in $\mathcal{S}$. Of course, L is *pure-injective* if every pure-exact sequence $0 \to L \to M \to N \to 0$ splits. Given any category $\mathcal{C}$ having a class of pure-injective sequences, we denote by $\operatorname{Pinj}\mathcal{C}$ the full subcategory of pure-injective objects and $\operatorname{Ind}\mathcal{C}$ denotes the isomorphism classes of indecomposable pure-injective objects in $\mathcal{C}$.

THEOREM 2.14. *Let $\mathcal{X}$ be a definable subcategory of* $\operatorname{Mod} R$. *The inclusion* $i\colon \mathcal{X} \to \operatorname{Mod} R$ *and the corresponding quotient functor* $q\colon \operatorname{Mod} R \to \mathcal{Z}$ *have the following properties:*

(1) *$q(M) = 0$ if and only if M lies in $\mathcal{X}$.*
(2) *i and q preserve direct limits, products and pure-exact sequences.*
(3) *i induces a fully faithful functor* $\operatorname{Pinj} \mathcal{X} \to \operatorname{Pinj} R$.
(4) *q induces a full and dense functor* $\operatorname{Pinj} R \to \operatorname{Pinj} \mathcal{Z}$. *If ϕ is a morphism in* $\operatorname{Pinj} R$, *then $q(\phi) = 0$ if and only if ϕ factors through an object in* $\operatorname{Pinj} \mathcal{X}$.
(5) *q induces a bijection* $\operatorname{Ind} R \setminus \operatorname{Ind} \mathcal{X} \to \operatorname{Ind} \mathcal{Z}$.

PROOF. We sketch the proof and refer the reader to [**52**] for details. Denote for any abelian category $\mathcal{C}$ by $\operatorname{Lex}(\mathcal{C}, \mathrm{Ab})$ the category of left exact functors $\mathcal{C} \to \mathrm{Ab}$. The sequence of exact functors $C(\mathcal{Z}) \xrightarrow{j} C(R) \xrightarrow{p} C(\mathcal{X})$ in combination with the equivalence $D(R) \to \operatorname{Lex}(C(R)^{\mathrm{op}}, \mathrm{Ab})$, $X \mapsto \operatorname{Hom}(-, X)|_{C(R)}$, induces the following commutative diagram of functors [**50**, Lemma 2.7]:

$$\begin{array}{ccccc} \mathcal{T} & \longrightarrow & D(R) & \xrightarrow{q_{\mathcal{X}}} & D(R)/\mathcal{T} \\ \downarrow\wr & & \downarrow\wr & & \downarrow\wr \\ \operatorname{Lex}(C(\mathcal{Z})^{\mathrm{op}}, \mathrm{Ab}) & \xrightarrow{j^*} & \operatorname{Lex}(C(R)^{\mathrm{op}}, \mathrm{Ab}) & \xrightarrow{p^*} & \operatorname{Lex}(C(\mathcal{X})^{\mathrm{op}}, \mathrm{Ab}) \end{array}$$

An analysis of the right half of the diagram provided the properties of the definable subcategory $\mathcal{X}$, and similar arguments can be used to prove the statements about the quotient functor $q\colon \operatorname{Mod} R \to \mathcal{Z}$. □

We continue with an example which illustrates the relation between a definable subcategory and its quotient category; it is taken from [**52**].

Let $\mathcal{C}$ be a full additive subcategory of $\operatorname{mod} R$ and suppose that $\mathcal{C}$ has split idempotents and is covariantly finite in $\operatorname{mod} R$. Denote by $\underline{\operatorname{mod}}_{\mathcal{C}} R$ the corresponding stable category, i.e. $\underline{\operatorname{mod}}_{\mathcal{C}} R$ has the same objects as $\operatorname{mod} R$ but $\underline{\operatorname{Hom}}_{\mathcal{C}}(X, Y)$ is $\operatorname{Hom}_R(X, Y)$ modulo the subgroup of morphisms which factor through an object in $\mathcal{C}$.

PROPOSITION 2.15. *The subcategory $\varinjlim \mathcal{C}$ is definable and the inclusion functor $\varinjlim \mathcal{C} \to \operatorname{Mod} R$ induces a commutative diagram of canonical functors*

$$\begin{array}{ccccc} \mathcal{C} & \longrightarrow & \operatorname{mod} R & \xrightarrow{p} & \underline{\operatorname{mod}}_{\mathcal{C}} R \\ \downarrow & & \downarrow & & \downarrow \\ \varinjlim \mathcal{C} & \longrightarrow & \operatorname{Mod} R & \xrightarrow{q} & \mathcal{Z} \end{array}$$

having the following properties:

(1) *p is the additive quotient functor for $\mathcal{C}$ and q is the definable quotient functor for $\varinjlim \mathcal{C}$.*
(2) *Each vertical functor $\mathcal{A} \to \mathcal{B}$ induces an equivalence between $\mathcal{A}$ and* $\operatorname{fp} \mathcal{B}$, *and each object in $\mathcal{B}$ is a direct limit of finitely presented objects.*
(3) $C(\varinjlim \mathcal{C}) \simeq \operatorname{fp}(\mathcal{C}, \mathrm{Ab})^{\mathrm{op}}$ *and* $C(\mathcal{Z}) \simeq \operatorname{fp}(\underline{\operatorname{mod}}_{\mathcal{C}} R, \mathrm{Ab})^{\mathrm{op}}$.

PROOF. For the definition of a covariantly finite subcategory and the fact that $\varinjlim \mathcal{C}$ is definable, see the discussion of left approximations. For the rest, see [**52**]. □

EXAMPLE 2.16. Let $\mathcal{C} = \operatorname{proj} R$ be the category of finitely generated projective R-modules. Then $\underline{\operatorname{mod}} R = \underline{\operatorname{mod}}_{\mathcal{C}} R$ is the *stable category* of finitely presented R-modules. Using the results from Theorem 2.14 and Proposition 2.15, it has been shown in [**49**] that two finite dimensional algebras have the same representation type if their stable module categories are equivalent. More precisely, $\underline{\operatorname{mod}} R \simeq \underline{\operatorname{mod}} S$ implies that R has tame representation type if and only if S has tame representation type.

2.5. Examples

We present a series of examples which illustrate the omnipresence of definable subcategories.

(1) Let $\mathcal{C}$ be any class of R-modules and denote by Φ the collection of maps ϕ in $\operatorname{mod} R$ such that $\operatorname{Hom}_R(\phi, M)$ is surjective for all M in $\mathcal{C}$. Then the full subcategory of Φ-injective R-modules is the smallest definable subcategory containing $\mathcal{C}$.

(2) Let $\mathcal{C}$ be a class of finitely presented R-modules. Then the R-modules M satisfying $\operatorname{Hom}_R(\mathcal{C}, M) = 0$ form a definable subcategory.

(3) Given a collection Φ of maps in $\operatorname{Mod} R$ we denote by $(\operatorname{Mod} R)^{\Phi}$ the full subcategory of Φ-*projective* R-modules M, i.e. each map $X \to Y$ in Φ induces a surjection $\operatorname{Hom}_R(M, X) \to \operatorname{Hom}_R(M, Y)$. Suppose now that R is an artin k-algebra. Then for each collection Φ of maps in $\operatorname{mod} R$ there are collections Φ' and Φ'' in $\operatorname{mod} R$ such that $(\operatorname{Mod} R)_{\Phi} = (\operatorname{Mod} R)^{\Phi'}$ and $(\operatorname{Mod} R)^{\Phi} = (\operatorname{Mod} R)_{\Phi''}$. This follows from the functorial isomorphism $\operatorname{Hom}_R(M, X) \simeq (M \otimes_R X^*)^*$ for arbitrary M and finitely presented X in $\operatorname{Mod} R$ ($N^* = \operatorname{Hom}_k(N, k)$ denotes the usual k-dual).

(4) Let $\mathcal{C}$ be a full additive subcategory of $\operatorname{mod} R$ and denote by $\varinjlim \mathcal{C}$ the full subcategory of R-modules which are direct limits of R-modules in $\mathcal{C}$. Then $\varinjlim \mathcal{C}$ is definable if and only if $\mathcal{C}$ is a covariantly finite subcategory of $\operatorname{mod} R$, see Proposition 3.11.

(5) Let $R = kG$ be the group algebra of a finite group G over a field k of prime characteristic, and let V be a closed homogeneous subvariety of the maximal ideal spectrum of the cohomology ring $H^*(G, k)$. Denoting by $\mathcal{C}_V$ the class of finitely generated R-modules whose variety is contained in V, the $\mathcal{C}_V$-local modules form a definable subcategory of $\operatorname{Mod} R$. Recall that a module M is $\mathcal{C}_V$*-local* if every map $X \to M$ with X in $\mathcal{C}_V$ factors through a projective module [**69**].

(6) Let R be a hereditary artin algebra. Recall from [**70**] that an R-module M is called *torsion-free* (respectively, *divisible*) if $\operatorname{Hom}_R(X, M) = 0$ (respectively, $\operatorname{Ext}^1_R(X, M) = 0$) for every regular R-module X. The torsion-free modules form a definable subcategory which is of the form $\varinjlim \mathcal{P}$ ($\mathcal{P}$ the category of finitely presented preprojective modules). Also the divisble modules form a definable subcategory.

(7) Suppose that M is an endofinite R-module, i.e. M is of finite length when regarded as an $\operatorname{End}_R(M)^{\mathrm{op}}$-module. Then $\operatorname{Add} M$ is a definable subcategory. Here $\operatorname{Add} M$ denotes the full subcategory of R-modules which are direct summands of coproducts of copies of M.

(8) Let R be a tame hereditary artin algebra. Then the torsion-free divisible R-modules in the sense of [**70**] form a definable subcategory which is of the form $\operatorname{Add} M$ for some endofinite R-module M.

(9) Let R be right coherent and denote by Φ the collection of all monos in $\operatorname{mod} R$. Then an R-module M is Φ-injective if and only if M is fp-injective, i.e.

$\mathrm{Ext}^1_R(-, M)$ vanishes on every finitely presented module. Note that fp-injective and injective R-modules coincide if and only if R is right noetherian, see Proposition A.11.

(10) Let R be left artinian. Then for each $n \geq 0$ the full subcategory of R-modules of projective dimension at most n is definable. In particular, the big finitistic dimension of R is the supremum of the projective dimensions of the modules in the Ziegler spectrum having finite projective dimension; it is finite if and only if the full subcategory of R-modules having finite projective dimension is definable [**53**].

(11) Let $\phi\colon X \to Y$ be a left almost split map in $\mathrm{Mod}\, R$. Then an R-module M is ϕ-injective if and only if M has no direct summand which is isomorphic to X.

(12) Let S be a subset of R. Then the S-divisible R-modules form a definable subcategory. Recall that an R-module M is *S-divisible* if $Ms = M$ for all $s \in S$.

(13) Let R be a semi-prime right noetherian ring. Then the torsion-free R-modules form a definable subcategory. Recall that an R-module M is *torsion-free* if $m \cdot r \neq 0$ for every $0 \neq m \in M$ and every regular $r \in R$.

CHAPTER 3

Left approximations

In this chapter we discuss two types of left approximations. The first type are the left almost split maps which have been introduced by Auslander and Reiten [**6**]. This concept is one of the most successful in modern algebra representation theory. The second type of left approximations is defined with respect to a class $\mathcal{C}$ of R-modules. This concept has been introduced by Auslander and Smalø [**7**], and independently by Enochs [**25**].

3.1. Left almost split maps

Recall from [**6**] that a map $\phi\colon M \to N$ in some additive category $\mathcal{C}$ is *left almost split* if ϕ is not a split mono and if every map $M \to L$ which is not a split mono factors through ϕ. It is important to observe that the endomorphism ring of M is local if there exists a left almost split map $M \to N$. In this section we discuss left almost split maps in $\operatorname{mod} R$ and $\operatorname{Mod} R$; they are closely related to the injective envelopes of simple objects in $C(R)$ and $D(R)$. We begin with a result which is due to Crawley-Boevey.

PROPOSITION 3.1. *The following are equivalent for a pure-injective R-module M:*

(1) *There is a left almost split map $M \to N$ in* $\operatorname{Mod} R$.
(2) *T_M is the injective envelope of a simple object in $D(R)$.*

PROOF. See [**18**, Theorem 2.3]. □

The description of the injective envelopes of the simples in $D(R)$ has an immediate consequence for the Ziegler spectrum.

COROLLARY 3.2. *The set of modules M in* $\operatorname{Ind} R$ *which admit a left almost split map $M \to N$ in* $\operatorname{Mod} R$ *form a dense subset of the Ziegler spectrum of R.*

PROOF. In view of Corollary 2.13 it suffices to show that for every non-zero $F \in C(R)$ there exists $M \in \operatorname{Ind} R$ with a left almost split map $M \to N$ such that $\operatorname{Hom}(F, T_M) \neq 0$. To this end choose a maximal subobject $G \subsetneq F$ and let T_M be the injective envelope of F/G. Clearly, $\operatorname{Hom}(F, T_M) \neq 0$. □

It also possible to characterize the existence of a left almost split map using only the modules in $\operatorname{Ind} R$. This generalizes a result of Auslander in [**4**] where it is shown that every finitely presented indecomposable module over an artin algebra satisfies condition (2) of the following theorem.

THEOREM 3.3. *The following are equivalent for a pure-injective R-module M:*

(1) *There is a left almost split map $M \to M'$ in* $\operatorname{Mod} R$.

(2) *For every product $N = \prod_{i\in I} N_i$ of indecomposable R-modules such that M is isomorphic to a direct summand of N there exists $i \in I$ such that $M \simeq N_i$.*
(3) *For every product $N = \prod_{i\in I} N_i$ of modules in* $\operatorname{Ind} R$ *such that M is isomorphic to a direct summand of N there exists $i \in I$ such that $M \simeq N_i$.*

PROOF. We shall work with the characterization of (1) from Proposition 3.1.

(1) $\Rightarrow$ (2) Suppose that T_M is the injective envelope of S in $D(R)$. If M is isomorphic to a direct summand of a product $N = \prod_{i\in I} N_i$ of indecomposable R-modules, then $\operatorname{Hom}(S, T_N) \neq 0$ and therefore $\operatorname{Hom}(S, T_{N_i}) \neq 0$ for some $i \in I$. Thus there is a non-zero morphism

$$S \longrightarrow T_M \longrightarrow T_N \longrightarrow T_{N_i}$$

and therefore $T_M \to T_{N_i}$ is a monomorphism since T_M is an injective envelope of S. It follows that $M \simeq N_i$ since M is pure-injective and N_i is indecomposable.

(2) $\Rightarrow$ (3) Trivial.

(3) $\Rightarrow$ (1) Suppose (3). Clearly, M needs to be indecomposable since every module is a pure submodule of some product of modules in $\operatorname{Ind} R$, for instance by Theorem 2.11. Consider for the set $\mathbf{U}$ of modules in $\operatorname{Ind} R$ which are not isomorphic to M the canonical map

$$\phi\colon M \longrightarrow \prod_{N\in\mathbf{U}} N^{\operatorname{Hom}_R(M,N)}$$

and let $F = \operatorname{Ker} T_\phi$. Assuming (3) we have $F \neq 0$ and find a finitely generated subobject $0 \neq G \subseteq F$. Taking a maximal subobject $H \subseteq G$ there exist an injective envelope $G/H \to T_L$ for some $L \in \operatorname{Ind} R$. The composition with $G \to G/H$ extends to a non-zero map $T_\psi\colon T_M \to T_L$ and ψ does not factor through ϕ. Therefore $M \simeq L$ and there exist a left almost split map starting in M since T_M is an injective envelope of G/H. □

We have the following analogue of Proposition 3.1 for left almost split maps in $\operatorname{mod} R$.

PROPOSITION 3.4. *For a finitely presented R-module M the following are equivalent:*

(1) *There is a left almost split map $M \to N$ in* $\operatorname{mod} R$.
(2) *T_M is the injective envelope of a simple object in $C(R)$.*

PROOF. We use the description of the injective objects in $C(R)$ which follows from Lemma 1.6.

(1) $\Rightarrow$ (2) Let $\phi\colon M \to N$ be left almost split in $\operatorname{mod} R$ and consider the kernel $F = \operatorname{Ker} T_\phi$ in $C(R)$. We have $F \neq 0$ since $M \to N$ is not a split mono. Now if F is not simple, say with proper non-zero subobject G, then T_M/G can be embedded in an injective object T_L in $C(R)$. The corresponding map $M \to L$ needs to factor through ϕ, a contradiction. Thus F is simple. A similar argument shows that $F \subseteq G$ for every non-zero subobject $G \subseteq T_M$. Thus T_M is the injective envelope of a simple object in $C(R)$.

(2) $\Rightarrow$ (1) Let S be a simple subobject of T_M and embed T_M/S into an injective object T_N in $C(R)$. It is easily checked that the corresponding map $M \to N$ is left almost split in $\operatorname{mod} R$. □

The existence of a left almost split map $M \to N$ for $M \in \operatorname{Ind} R$ is closely related to the property of M to be isolated in the Ziegler spectrum. Let us call M *Ziegler-isolated* if $\{M\}$ is a Ziegler-open subset.

PROPOSITION 3.5. *Let $M \in \operatorname{Ind} R$.*

(1) *If M is Ziegler-isolated, then there exist a left almost split map $M \to N$ in* $\operatorname{Mod} R$, *i.e. T_M is the injective envelope of a simple object in $D(R)$.*
(2) *If T_M is the injective envelope of a finitely presented simple object in $D(R)$, then M is Ziegler-isolated.*

PROOF. (1) If M is Ziegler-isolated, then there exists $0 \neq F \in C(R)$ such that $\operatorname{Hom}(F, T_N) = 0$ for all $N \in \operatorname{Ind} R$ different from M. Choosing a maximal subobject $G \subseteq F$ it is clear that T_M is the injective envelope of F/G.

(2) Suppose T_M is the injective envelope of a finitely presented simple object S. Clearly, $\operatorname{Hom}(S, T_N) = 0$ for all $N \in \operatorname{Ind} R$ different from M and therefore M is Ziegler-isolated. □

We combine now the existence of left almost split maps in $\operatorname{mod} R$ and $\operatorname{Mod} R$ and obtain in this way another criterion for a module in the Ziegler spectrum to be isolated.

THEOREM 3.6. *Let $M \in \operatorname{Mod} R$ be the pure-injective envelope of $N \in \operatorname{mod} R$. Then the following are equivalent:*

(1) *There exist a left almost split map $M \to M'$ in* $\operatorname{Mod} R$.
(2) *There exist a left almost split map $N \to N'$ in* $\operatorname{mod} R$.

Moreover, in this case M is Ziegler-isolated in $\operatorname{Ind} R$.

PROOF. Let $N \to M$ be the pure-injective envelope of N, i.e. the map $T_N \to T_M$ is an injective envelope in $D(R)$.

(1) $\Rightarrow$ (2) Assuming (1) it follows from Proposition 3.1 that T_M is the injective envelope of a simple object S in $D(R)$ which needs to be finitely presented since it is a finitely generated subobject of T_N. Therefore T_N is the injective envelope of S in $C(R)$ and the existence of a left almost split map $N \to N'$ now follows from Proposition 3.4.

(2) $\Rightarrow$ (1) We use again Proposition 3.1 and Proposition 3.4. Assuming (2) it follows that T_N is an injective envelope of a simple object S in $C(R)$. Therefore T_M is injective envelope of S in $D(R)$ since every non-zero subobject F of T_M has a finitely presented non-zero subobject $G \subseteq F$ which is contained in T_N and contains therefore S. Thus there is a left almost split map $M \to M'$.

The fact that M is Ziegler-isolated is a consequence of Proposition 3.5. □

EXAMPLE 3.7. Let R be an artin algebra. Then a module $M \in \operatorname{Ind} R$ is Ziegler-isolated if and only if M is finitely presented if and only if there exists a left almost split map $M \to N$.

3.2. Left approximations

Let $\mathcal{A}$ be any category and $\mathcal{C}$ be any class of objects in $\mathcal{A}$. Following **[7]**, a morphism $M \to N$ in $\mathcal{A}$ is called a *left $\mathcal{C}$-approximation* of M provided that N belongs to $\mathcal{C}$ and the induced map $\operatorname{Hom}(N, C) \to \operatorname{Hom}(M, C)$ is surjective for every C in $\mathcal{C}$. If every object in $\mathcal{A}$ has a left $\mathcal{C}$-approximation, then $\mathcal{C}$ is said to be *covariantly finite* in $\mathcal{A}$. In this section we want to study left $\mathcal{C}$-approximations of modules over a ring R where $\mathcal{C}$ is the class of modules belonging to a definable subcategory of $\operatorname{Mod} R$.

We begin our discussion with a criterion for a class of modules to be covariantly finite. Suppose that $\mathcal{C}$ is a class of R-modules which is closed under direct summands. It is easily seen that $\mathcal{C}$ needs to be closed under taking products if $\mathcal{C}$ is covariantly finite in $\operatorname{Mod} R$. The converse is also true if $\mathcal{C}$ is closed under pure submodules [**68**].

LEMMA 3.8. *Let $\mathcal{C}$ be any class of R-modules which is closed under pure submodules. Then the following are equivalent:*

(1) *Every R-module has a left $\mathcal{C}$-approximation.*
(2) *$\mathcal{C}$ is closed under taking products.*

PROOF. (1) $\Rightarrow$ (2) Let $M = \prod_{i\in I} M_i$ be a product of modules in $\mathcal{C}$ and let $\phi\colon M \to N$ be a left $\mathcal{C}$-approximation. Every projection $M \to M_i$ factors through ϕ and therefore ϕ is a split monomorphism. It follows that M belongs to $\mathcal{C}$ since $\mathcal{C}$ is closed under direct summands by assumption.

(2) $\Rightarrow$ (1) It is well-known that every submodule M of a module N is contained in a pure submodule M' of N with $\operatorname{card} M' \leq \sup(\operatorname{card} M, \operatorname{card} R, \aleph_0)$, see [**45**]. Therefore consider for every R-module M the set $\mathcal{C}_M$ of isomorphism classes of modules N in $\mathcal{C}$ with $\operatorname{card} N \leq \sup(\operatorname{card} M, \operatorname{card} R, \aleph_0)$. It follows that the canonical map $M \to \prod_{N\in\mathcal{C}_M} N^{\operatorname{Hom}_R(M,N)}$ is a left $\mathcal{C}$-approximation. □

We continue with two other lemmas.

LEMMA 3.9. *Let $\mathcal{C}$ be a full additive subcategory of* $\operatorname{mod} R$. *Then an R-module M is a direct limit of objects in $\mathcal{C}$ if and only if every morphism $X \to M$ with $X \in \operatorname{mod} R$ factors through a module in $\mathcal{C}$.*

PROOF. See [**20**, Lemma 4.1]. □

LEMMA 3.10. *Let $\mathcal{C}$ be a full additive subcategory of* $\operatorname{mod} R$. *Then $\varinjlim \mathcal{C}$ is closed under pure submodules.*

PROOF. Let $0 \to L \to M \to N \to 0$ be a pure-exact sequence with $M \in \varinjlim \mathcal{C}$, and let $X \to L$ be a map with $X \in \operatorname{mod} R$. We apply the criterion from Lemma 3.9. The composition $X \to L \to M$ factors through some Y in $\mathcal{C}$ and we obtain therefore the following commutative diagram with exact rows

$$\begin{array}{ccccccccc} & & X & \longrightarrow & Y & \longrightarrow & Z & \longrightarrow & 0 \\ & & \downarrow & & \downarrow & & \downarrow & & \\ 0 & \longrightarrow & L & \longrightarrow & M & \longrightarrow & N & \longrightarrow & 0 \end{array}$$

The map $Z \to N$ factors through M since $L \to M$ is pure, and therefore $X \to L$ factors through Y. Thus L belongs to $\varinjlim \mathcal{C}$. □

We are now in a position to characterize the definable subcategories of $\operatorname{Mod} R$ which are of the form $\varinjlim \mathcal{C}$ for some additive subcategory $\mathcal{C}$ of $\operatorname{mod} R$.

PROPOSITION 3.11. *Let $\mathcal{C}$ be a full additive subcategory of* $\operatorname{mod} R$. *Then the following are equivalent for $\mathcal{X} = \varinjlim \mathcal{C}$:*

(1) *$\mathcal{X}$ is definable.*
(2) *Every R-module has a left $\mathcal{X}$-approximation.*
(3) *Every finitely presented R-module has a left $\mathcal{X}$-approximation.*

(4) *Every finitely presented R-module has a left $\mathcal{C}$-approximation.*

PROOF. (1) $\Rightarrow$ (2) $\mathcal{X}$ is closed under products and pure submodules by Theorem 2.1 and therefore every R-module has a left $\mathcal{X}$-approximation by Lemma 3.8.

(2) $\Rightarrow$ (3) Trivial.

(3) $\Rightarrow$ (4) Let $X \to M$ be a left $\mathcal{X}$-approximation for $X \in \operatorname{mod} R$. The map factors through some $N \in \mathcal{C}$ by Lemma 3.9. The corresponding map $X \to N$ is a left $\mathcal{C}$-approximation for X.

(4) $\Rightarrow$ (1) $\mathcal{X}$ is closed under taking direct limits by construction and closed under pure submodules by Lemma 3.10. In [**20**], it is shown that $\mathcal{X}$ is also closed under taking products. In fact, let $M = \prod_{i \in I} M_i$ be a product of modules in $\mathcal{X}$ and let $\phi\colon X \to M$ be a map with X finitely presented. We choose a left $\mathcal{C}$-approximation $\psi\colon X \to N$. Each component $X \to M_i$ of ϕ factors through some module in $\mathcal{C}$, by Lemma, and factors therefore also through ψ. It follows that ϕ factors through ψ and therefore M belongs to $\varinjlim \mathcal{C}$ by Lemma 3.9. Now it follows from Theorem 2.1 that $\mathcal{X}$ is definable. □

The following result summarizes our discussion of left approximations for definable subcategories.

THEOREM 3.12. *Let $\mathcal{X}$ be a definable subcategory of* $\operatorname{Mod} R$. *Then every R-module has a left $\mathcal{X}$-appropximation. Moreover, the following are equivalent:*

(1) *Every finitely presented R-module has a left $\mathcal{X}$-approximation which is finitely presented.*
(2) *There is a full additive subcategory $\mathcal{C}$ of* $\operatorname{mod} R$ *such that $\mathcal{X} = \varinjlim \mathcal{C}$.*

PROOF. $\mathcal{X}$ is closed under taking products and pure submodules by Theorem 2.1, and therefore every R-module has a left $\mathcal{X}$-approximation by Lemma 3.8.

(1) $\Rightarrow$ (2) Let $\mathcal{C} = \mathcal{X} \cap \operatorname{mod} R$. We claim that $\mathcal{X} = \varinjlim \mathcal{C}$. One inclusion is clear since $\mathcal{X}$ is closed under taking direct limits. Therefore let $M \in \mathcal{X}$ and choose a map $\phi\colon X \to M$ with $X \in \operatorname{mod} R$. Condition (1) implies that ϕ factors through some module in $\mathcal{C}$ and therefore $M \in \varinjlim \mathcal{C}$ by Lemma 3.9. Thus $\mathcal{X} = \varinjlim \mathcal{C}$ and (2) follows.

(2) $\Rightarrow$ (1) Let X be a finitely presented R-module and let $\phi\colon X \to M$ be a left $\mathcal{X}$-approximation. If $\mathcal{X} = \varinjlim \mathcal{C}$, then ϕ factors through a map $\psi\colon X \to N$ with $N \in \mathcal{C}$ by Lemma 3.9. Clearly, ψ is a left $\mathcal{X}$-approximation. □

3.3. Minimal left approximations

Recall that a morphism $\phi\colon M \to N$ in any category is *left minimal* provided that any endomorphism ψ of N satisfying $\psi \circ \phi = \phi$ is an isomorphism. In this section we shall discuss criteria for the existence of left approximations which are left minimal. Our strategy is to reduce the problem of finding minimal left approximations to the existence of injective envelopes in the category $D(R)$.

PROPOSITION 3.13. *Let $\phi\colon M \to N$ be a map in* $\operatorname{Mod} R$ *and suppose that N is pure-injective. Then there exists a decomposition*

$$\phi = (\phi', \phi'')\colon M \longrightarrow N' \coprod N'' = N$$

such that ϕ' is left minimal and $\phi'' = 0$.

PROOF. Let $\psi = (\psi', \psi'') \colon \operatorname{Im} T_\phi \to T_{N'} \coprod T_{N''} = T_N$ be a decomposition such that ψ' is an injective envelope of $\operatorname{Im} T_\phi$ and $\psi'' = 0$. Clearly, this decomposition induces a decomposition $\phi = (\phi', \phi'') \colon M \to N' \coprod N'' = N$ of ϕ. The map ϕ' is left minimal since ψ' is an injective envelope, and $\phi'' = 0$ since $\psi'' = 0$. □

A decomposition $\phi = (\phi', \phi'') \colon M \to N' \coprod N'' = N$ of a map ϕ such that ϕ' is left minimal and $\phi'' = 0$ is unique up to isomorphism. The map ϕ' is sometimes called the *minimal version* of ϕ. Many existence results for minimal left approximations can be derived from the following theorem.

THEOREM 3.14. *Let $\mathcal{C}$ be any class of modules which is closed under direct summands. Let M be an R-module and suppose that M has a left $\mathcal{C}$-approximation which is pure-injective. Then M has a minimal left $\mathcal{C}$-approximation.*

PROOF. Apply the preceding proposition. □

We present various consequences of the preceding theorem.

COROLLARY 3.15. *Let $\mathcal{C}$ be any class of pure-injective R-modules which is closed under direct summands. Then the following are equivalent:*

(1) *Every R-module has a left $\mathcal{C}$-approximation.*
(2) *Every R-module has a minimal left $\mathcal{C}$-approximation.*
(3) *$\mathcal{C}$ is closed under taking products.*

PROOF. (1) $\Rightarrow$ (2) Use Theorem 3.14.

(2) $\Rightarrow$ (3) Use the argument in the proof of Lemma 3.8.

(3) $\Rightarrow$ (1) Let $\mathcal{D}$ be the class of pure submodules of modules in $\mathcal{C}$. Then every R-module M has a left $\mathcal{D}$-approximation $M \to N$ by Lemma 3.8. Suppose that N is the pure submodule of $N' \in \mathcal{C}$. It is clear that the composition $M \to N'$ is a left $\mathcal{C}$-approximation. □

EXAMPLE 3.16. (1) Let $\mathcal{C}$ be the class of all pure-injective R-modules. Then a morphism $\phi \colon M \to N$ is a minimal left $\mathcal{C}$-approximation if and only if ϕ is a pure-injective envelope of M (i.e. N is pure-injective and any map $\psi \colon N \to N'$ is a pure monomorphism if and only if $\psi \circ \phi$ is a pure monomorphism). In deed, the canonical map $\phi_M \colon M \to M^{**}$ (where $M^* = \operatorname{Hom}_{\mathbb{Z}}(M, \mathbb{Q}/\mathbb{Z})$) is a left $\mathcal{C}$-approximation since M^{**} is pure-injective and ϕ_M is a pure monomorphism. Therefore a minimal left $\mathcal{C}$-approximation $M \to N$ is a pure monomorphism, hence a pure-injective envelope. Conversely, if $\phi \colon M \to N$ is a pure-injective envelope, then any endomorphism ψ of N satisfying $\phi = \psi \circ \phi$ needs to be a pure-monomorphism, hence a split monomorphism, and therefore an isomorphism. The fact that every module has a pure-injective envelope is a classical result of Kiełpiński [**45**].

(2) Let $\mathcal{C}$ be the class of all injective R-modules. Then a morphism $\phi \colon M \to N$ is a minimal left $\mathcal{C}$-approximation if and only if ϕ is an injective envelope of M.

COROLLARY 3.17. *Let $\mathcal{X}$ be a definable subcategory of* $\operatorname{Mod} R$ *and denote by* $\operatorname{Pinj} \mathcal{X}$ *the class of pure-injective modules in $\mathcal{X}$. Then every R-module has a minimal left* $\operatorname{Pinj} \mathcal{X}$*-approximation.*

PROOF. $\operatorname{Pinj} \mathcal{X}$ is closed under taking products and direct summands. Therefore the assertion follows from Corollary 3.15. □

COROLLARY 3.18. *Let $\mathcal{X}$ be a definable subcategory of* $\operatorname{Mod} R$ *which is generated by a Σ-pure-injective R-module. Then every R-module has a minimal left $\mathcal{X}$-approximation.*

PROOF. The assumption on $\mathcal{X}$ implies that every module in $\mathcal{X}$ is pure-injective. This follows from Corollary 2.7; see also Lemma 6.5. The existence of minimal left $\mathcal{X}$-approximations is now a consequence of Corollary 3.15 since $\mathcal{X}$ is closed under taking products. □

EXAMPLE 3.19. Let R be a left coherent and right perfect ring. Then the flat R-modules form a definable subcategory consisting of Σ-pure-injective R-modules. Therefore every R-module has a so-called flat envelope [**25, 1**].

Given a definable subcategory $\mathcal{X}$ of $\operatorname{Mod} R$ it seems to be an open question under which conditions every R-module possesses left $\mathcal{X}$-approximation which is minimal. A sufficent condition has been formulated in Corollary 3.18. Another sufficent condition is the property of $\mathcal{X}$ that the inclusion functor $\mathcal{X} \to \operatorname{Mod} R$ has a left adjoint. We end this section with another class of examples.

EXAMPLE 3.20. Let $R = kG$ be the group algebra of a finite group G over a commutative field k of prime characteristic. Let $\mathcal{C}$ be an épaisse subcategory of the stable category $\underline{\operatorname{mod}}\, R$, i.e. a full triangulated subcategory that is closed under taking direct summands. Then the full subcategory $\mathcal{C}'$ of R-modules M such that $\underline{\operatorname{Hom}}_R(\mathcal{C}, M) = 0$ is definable. It follows from Rickard's work that every R-module M has a minimal left $\mathcal{C}'$-approximation $M \to F_{\mathcal{C}}(M)$, see [**69**]. If $\mathcal{C}$ is a tensor-ideal subcategory, then the minimal left $\mathcal{C}'$-approximation $F_{\mathcal{C}}(k)$ of the trivial R-module k is an idempotent module in the sense that $F_{\mathcal{C}}(k) \otimes_k F_{\mathcal{C}}(k) \simeq F_{\mathcal{C}}(k)$ in the stable category $\underline{\operatorname{Mod}}\, R$. Moreover, $F_{\mathcal{C}}(M) \simeq F_{\mathcal{C}}(k) \otimes_k M$ for every R-module M.

CHAPTER 4

Duality

The duality between $C(R)$ and $C(R^{\text{op}})$ induces a relation between R-modules and R^{op}-modules. The basic idea is to compare for two modules $M \in \operatorname{Mod} R$ and $N \in \operatorname{Mod} R^{\text{op}}$ the kernels of the corresponding exact functors $E_M \colon C(R) \to \text{Ab}$ and $E_N \colon C(R^{\text{op}}) \to \text{Ab}$. This approach leads to a duality map $M \mapsto M^\vee$ between certain classes of right and left R-modules which extends, for example, the usual k-duality between finite dimensional modules in case R is an algebra over a commutative field k.

4.1. Purely equivalent and purely opposed modules

In this section we discuss briefly two relations which are defined on the modules over R and its opposite R^{op}. This material is taken from [**48**]. We denote for every R-module M by $\mathcal{S}_M$ the Serre subcategory of $C(R)$ consisting of all F in $C(R)$ such that $\operatorname{Hom}(F, T_M) = 0$. Recall from Lemma 1.6 that there exists a canonical duality between $C(R)$ and $C(R^{\text{op}})$ which sends $F = \operatorname{Coker} H_\phi$ to $F^\vee = \operatorname{Ker} T_\phi$ for every map ϕ in $\operatorname{mod} R$ or $\operatorname{mod} R^{\text{op}}$.

LEMMA 4.1. *Let M and N be two R-modules. Then the following are equivalent:*

(1) *Any map ϕ in* $\operatorname{mod} R$ *induces an epi* $\operatorname{Hom}_R(\phi, M)$ *if and only if ϕ induces an epi* $\operatorname{Hom}_R(\phi, N)$.
(2) *Any map ψ in* $\operatorname{mod} R^{\text{op}}$ *induces a mono $M \otimes_R \psi$ if and only if ψ induces a mono $N \otimes_R \psi$.*
(3) $\mathcal{S}_M = \mathcal{S}_N$.

PROOF. Given any map ϕ in $\operatorname{mod} R$, there is ψ in $\operatorname{mod} R^{\text{op}}$ with $\operatorname{Coker} H_\phi = \operatorname{Ker} T_\psi$. Conversely, given any map ψ in $\operatorname{mod} R^{\text{op}}$, there is ϕ in $\operatorname{mod} R$ with $\operatorname{Coker} H_\phi = \operatorname{Ker} T_\psi$. This is a consequence of the duality between $C(R)$ and $C(R^{\text{op}})$ which sends $\operatorname{Coker} H_\phi$ to $\operatorname{Ker} T_\phi$. The assertion is an immediate consequence of this fact. □

Two R-modules M and N are called *purely equivalent* if they satisfy the equivalent conditions (1) - (3) of the preceding lemma.

LEMMA 4.2. *Let $M \in \operatorname{Mod} R$ and $N \in \operatorname{Mod} R^{\text{op}}$. Then the following are equivalent:*

(1) *Any map ϕ in* $\operatorname{mod} R$ *induces an epi* $\operatorname{Hom}_R(\phi, M)$ *if and only if ϕ induces a mono $\phi \otimes_R N$.*
(2) *Any map ψ in* $\operatorname{mod} R^{\text{op}}$ *induces an epi* $\operatorname{Hom}_{R^{\text{op}}}(\psi, N)$ *if and only if ψ induces a mono $M \otimes_R \psi$.*
(3) $\mathcal{S}_N = (\mathcal{S}_M)^\vee$.
(4) $\mathcal{S}_M = (\mathcal{S}_N)^\vee$.

PROOF. Use the same argument as in the proof of Lemma 4.1. □

Two modules $M \in \operatorname{Mod} R$ and $N \in \operatorname{Mod} R^{\mathrm{op}}$ are called *purely opposed* if they satisfy the equivalent conditions (1) - (4) of the preceding lemma.

LEMMA 4.3. *If M is an R-S-bimodule and $E \in \operatorname{Mod} S$ is an injective cogenerator, then the R-module M and the R^{op}-module $\operatorname{Hom}_S(M, E)$ are purely opposed.*

PROOF. If E is any injective S-module, then there is a well-known isomorphism

$$X \otimes_R \operatorname{Hom}_S(M, E) \longrightarrow \operatorname{Hom}_S(\operatorname{Hom}_R(X, M), E)$$

for all $X \in \operatorname{mod} R$ which is functorial in X. Taking a map ϕ in $\operatorname{mod} R$ it follows that $\phi \otimes_R \operatorname{Hom}_S(M, E)$ is a mono if and only if $\operatorname{Hom}_R(\phi, M)$ is an epi provided that E cogenerates $\operatorname{Mod} S$. Thus M and $\operatorname{Hom}_S(M, E)$ are purely opposed by Lemma 4.2. □

Suppose that R is a k-algebra where k denotes any commutative ring, and let E be an injective cogenerator for $\operatorname{Mod} k$. For instance, one could take $k = \mathbb{Z}$ and $E = \mathbb{Q}/\mathbb{Z}$. We obtain a functor

$$\operatorname{Mod} k \longrightarrow \operatorname{Mod} k, \quad M \mapsto M^* = \operatorname{Hom}_k(M, E)$$

which induces contravariant functors between $\operatorname{Mod} R$ and $\operatorname{Mod} R^{\mathrm{op}}$.

LEMMA 4.4. *An R-module M is purely opposed to M^* and purely equivalent to M^{**}.*

PROOF. Use the preceding lemmas. □

An advantage of passing from a module M to its dual M^* is the fact that M^* is always pure-injective. We formulate this well-known fact as follows.

LEMMA 4.5. *Let M be an R-S-bimodule. If $E \in \operatorname{Mod} S$ is injective, then $\operatorname{Hom}_S(M, E)$ is a pure injective R^{op}-module.*

PROOF. See [**5**, Proposition 10.1]. □

Our discussion of purely equivalent and purely opposed modules has an interesting consequence.

PROPOSITION 4.6. *A module is purely equivalent to its pure-injective envelope.*

PROOF. It is well-known that the canonical map $M \to M^{**}$ which sends $x \in M$ to $\operatorname{Hom}_k(M, E) \to E$, $\phi \to \phi(x)$, is a pure mono. The module M^{**} is pure-injective by Lemma 4.5 and purely equivalent to M by Lemma 4.4. Given any composition of pure monos $M \to N \to M^{**}$ it follows that M is purely equivalent to N. In particular, the pure-injective envelope of M which is a direct summand of M^{**} containing M is purely equivalent to M. □

4.2. The dual of a definable subcategory

A definable subcategory $\mathcal{X}$ of $\operatorname{Mod} R$ is determined by the corresponding Serre subcategory $\mathcal{S}_{\mathcal{X}}$ of $C(R)$. Using the duality ${}^{\vee}\colon C(R) \to C(R^{\mathrm{op}})$, we define a definable subcategory $\mathcal{X}^{\vee}$ of $\operatorname{Mod} R^{\mathrm{op}}$ by $\mathcal{S}_{\mathcal{X}^{\vee}} = (\mathcal{S}_{\mathcal{X}})^{\vee}$.

THEOREM 4.7. *The assignment $\mathcal{X} \mapsto \mathcal{X}^{\vee}$ gives a bijection between the definable subcategories of $\operatorname{Mod} R$ and the definable subcategories of $\operatorname{Mod} R^{\mathrm{op}}$.*

PROOF. Recall from Corollary 2.3 that the map $\mathcal{X} \mapsto \mathcal{S}_{\mathcal{X}}$ defines a bijection between the definable subcategories of $\operatorname{Mod} R$ and the Serre subcategories of $C(R)$. The map $\mathcal{S} \mapsto \mathcal{S}^\vee$ defines a bijection between the Serre subcategories of $C(R)$ and the Serre subcategories of $C(R^{\mathrm{op}})$. Combining these bijections the assertion follows. □

We include the following description of the dual $\mathcal{X}^\vee$ of a definable subcategory $\mathcal{X}$.

PROPOSITION 4.8. *Let $\mathcal{X}$ be a definable subcategory of* $\operatorname{Mod} R$. *Then an R^{op}-module M belongs to $\mathcal{X}^\vee$ if and only if M is purely opposed to some module in $\mathcal{X}$.*

PROOF. The assertion is a formal consequence of the definitions which are involved. □

EXAMPLE 4.9. Let Φ be the collection of all monos in $\operatorname{mod} R$ and let $\mathcal{X} = (\operatorname{Mod} R)_\Phi$. Then $\mathcal{X}^\vee$ is the definable subcategory of $\operatorname{Mod} R^{\mathrm{op}}$ which is generated by the finitely generated projective R^{op}-modules. If R is right coherent, then $\mathcal{X}$ is the category of fp-injective R-modules and $\mathcal{X}^\vee$ is the category of flat R^{op}-modules.

4.3. Pure-reflexive modules

Let R be a k-algebra over some commutative ring k. We fix an injective cogenerator E for $\operatorname{Mod} k$ and obtain a functor

$$\operatorname{Mod} k \longrightarrow \operatorname{Mod} k, \quad M \mapsto M^* = \operatorname{Hom}_k(M, E)$$

which induces contravariant functors between $\operatorname{Mod} R$ and $\operatorname{Mod} R^{\mathrm{op}}$. In this section we introduce a class of R-modules which we call pure-reflexive. We denote by $\operatorname{Ref} R$ the isomorphism classes of pure-reflexive R-modules and establish a bijection $M \mapsto M^\vee$ between $\operatorname{Ref} R$ and $\operatorname{Ref} R^{\mathrm{op}}$ which has the following properties:

(D1) $(M \coprod N)^\vee = M^\vee \coprod N^\vee$;
(D2) $M^{\vee\vee} = M$;
(D3) $M^\vee$ is a direct summand of M^*.

This duality between right and left R-modules covers Herzog's elementary duality [**39**], Crawley-Boevey's duality for endofinite modules [**18**], and the duality studied in [**48**]. In fact, the approach presented here is based to a large extent on work in [**48**].

We start our discussion with the definition of a pure-reflexive module. An indecomposable pure-injective R-module M with $S = \operatorname{End}_R(M)^{\mathrm{op}}$ is said to be *pure-reflexive* if there exists a map $X \to Y$ in $\operatorname{mod} R$ such that the cokernel of the induced map $\operatorname{Hom}_R(Y, M) \to \operatorname{Hom}_R(X, M)$ is a simple S-module. An arbitrary R-module M is called *pure-reflexive* if M is the pure-injective envelope of a coproduct $\coprod_{i \in I} M_i$ of pure-reflexive modules in $\operatorname{Ind} R$. We need a characterization of the indecomposable pure-reflexive R-modules. To this end fix an indecomposable pure-injective R-module M and let $S = \operatorname{End}_R(M)^{\mathrm{op}}$. We consider the exact functor

$$E_M \colon C(R) \longrightarrow \operatorname{Mod} S, \quad F \mapsto \operatorname{Hom}(F, T_M)$$

where S acts on $E_M(F)$ via the isomorphism $S \simeq \operatorname{End}(T_M)^{\mathrm{op}}$, and we denote by $\mathcal{S}_M$ the kernel of E_M which is a Serre subcategory of $C(R)$.

LEMMA 4.10. *The following are equivalent:*

(1) *M is pure-reflexive.*
(2) *The quotient category $C(R)/\mathcal{S}_M$ contains a simple object.*

For the proof of this lemma we need the following observation.

LEMMA 4.11. *Let N be the injective envelope of a simple object X in any abelian category. Then* $\operatorname{Hom}(X,N)$ *is a simple* $\operatorname{End}(N)^{\mathrm{op}}$*-module.*

PROOF. The inclusion $X \to N$ induces an epi $\operatorname{Hom}(N,N) \to \operatorname{Hom}(X,N)$ and its kernel is precisely the maximal ideal of all endomorphisms ϕ of N such that $\phi(X)=0$. □

PROOF OF LEMMA 4.10. Let $F = \operatorname{Ker} T_\phi$ be an object in $C(R)$. We shall use the S-linear isomorphism

$$\operatorname{Hom}(F, T_M) \simeq \operatorname{Coker} \operatorname{Hom}_R(\phi, M).$$

(1) $\Rightarrow$ (2) If $\operatorname{Coker}\operatorname{Hom}_R(\phi, M)$ is a simple S-module, then $\operatorname{Hom}(F,T_M)$ is a simple object in $C(R)/\mathcal{S}_M$ since E_M induces a functor $C(R)/\mathcal{S}_M \to \operatorname{Mod} S$ which is faithful and exact.

(2) $\Rightarrow$ (1) Let $\mathcal{T} = \varinjlim \mathcal{S}_M$. It follows from Proposition A.5 and Lemma 6.1 that the quotient category $D(R)/\mathcal{T}$ is locally coherent with

$$\operatorname{fp}(D(R)/\mathcal{T}) \simeq C(R)/\mathcal{S}_M.$$

If $\operatorname{Hom}(F,T_M)$ is a simple object in $C(R)/\mathcal{S}_M$, then F is also simple in $D(R)/\mathcal{T}$. It follows that any non-zero map $F \to T_M$ becomes an injective envelope in $D(R)/\mathcal{T}$, and $\operatorname{Hom}_{D(R)/\mathcal{T}}(F,T_M)$ can be identified with the S-module $\operatorname{Coker}\operatorname{Hom}_R(\phi,M)$ since T_M is $\mathcal{T}$-closed. We conclude from the preceding lemma that this S-module is simple. □

The following result is taken from [**48**].

PROPOSITION 4.12. *Let $M \in \operatorname{Ind} R$ and suppose that M is pure-reflexive. Then there exist a unique module in* $\operatorname{Ind} R^{\mathrm{op}}$ *which is purely opposed to M. This module is denoted by $M^\vee$; it is pure-reflexive and $M^{\vee\vee} = M$.*

PROOF. We use the canonical duality ${}^\vee\colon C(R) \to C(R^{\mathrm{op}})$. Let $\mathcal{S} = \mathcal{S}_M$ and $\mathcal{T} = \varinjlim \mathcal{S}$ be the corresponding localizing subcategory of $D(R)$. Note that ${}^\vee$ induces a duality $C(R)/\mathcal{S} \to C(R^{\mathrm{op}})/\mathcal{S}^\vee$ and let $\mathcal{T}^\vee = \varinjlim \mathcal{S}^\vee$. The assumption on M says that the category $C(R)/\mathcal{S}_M$ has a unique simple object S and therefore $C(R^{\mathrm{op}})/\mathcal{S}^\vee$ has also a unique simple object $S^\vee$. We view S as an object in $D(R)/\mathcal{T}$, and the section functor $D(R)/\mathcal{T} \to D(R)$ identifies the injective envelope of S in $D(R)/\mathcal{T}$ with T_M. Analogously, the section functor $D(R^{\mathrm{op}})/\mathcal{T}^\vee \to D(R^{\mathrm{op}})$ identifies the injective envelope of $S^\vee$ in $D(R^{\mathrm{op}})/\mathcal{T}^\vee$ with T_N for some $N \in \operatorname{Ind} R^{\mathrm{op}}$. We leave it to the reader to check that $\mathcal{S}_N = \mathcal{S}^\vee$, and therefore N is purely opposed to M. Moreover, N is unique since the injective envelope of $S^\vee$ is unique up to isomorphism. □

We proceed with some notation. Let M_i, $i \in I$, be a family of R-modules. Then we denote by $\widehat{\coprod}_{i\in I} M_i$ the pure-injective envelope of the coproduct $\coprod_{i\in I} M_i$. We shall need the following property of $\widehat{\coprod}$.

LEMMA 4.13. *Let M_i, $i \in I$, be a family of R-modules. If $I = \bigcup_{j\in J} I_j$ is a disjoint union, then*

$$\widehat{\coprod_{i\in I}} M_i = \widehat{\coprod_{j\in J}}(\widehat{\coprod_{i\in I_j}} M_i).$$

PROOF. A map $\phi\colon M \to N$ is a pure-injective envelope if and only if $T_M \to T_N$ is an injective envelope in $D(R)$, i.e. T_N is injective and T_ϕ is an essential monomorphism. Any coproduct of essential monomorphism is again essential, and therefore the map

$$\coprod_{i\in I} M_i = \coprod_{j\in J}(\coprod_{i\in I_j} M_i) \longrightarrow \coprod_{j\in J}(\widehat{\coprod_{i\in I_j}} M_i)$$

extends to an isomorphism between the pure-injective envelopes of $\coprod_{i\in I} M_i$ and $\coprod_{j\in J}(\widehat{\coprod}_{i\in I_j} M_i)$. □

We now define the map $M \mapsto M^\vee$ on the class $\operatorname{Ref} R$ of isomorphism classes of pure-reflexive R-modules. If $M = \widehat{\coprod}_{i\in I} M_i$ is pure-reflexive with $M_i \in \operatorname{Ind} R$ for all i, then we define $M^\vee = \widehat{\coprod}_{i\in I} M_i^\vee$ where $M_i^\vee$ is taken from Proposition 4.12. Note that $M^\vee$ is well-defined up to isomorphism by the Krull-Remak-Schmidt Theorem; see for instance [**27**]. We mention also that $\operatorname{Ref} R$ is closed under taking direct summands, finite coproducts, and arbitrary coproducts of the form $\widehat{\coprod}$. The pure-reflexive modules include all Σ-pure-injective modules. More generally, every module M with $\operatorname{KGdim} M < \infty$ is pure-reflexive; see Corollary 7.10.

THEOREM 4.14. *The assignment $M \mapsto M^\vee$ defines a bijection between* $\operatorname{Ref} R$ *and* $\operatorname{Ref} R^{\mathrm{op}}$ *which has the following properties:*

(1) $(\widehat{\coprod}_{i\in I} M_i)^\vee = \widehat{\coprod}_{i\in I} M_i^\vee$ *for every family* $(M_i)_{i\in I}$ *in* $\operatorname{Ref} R$. *In particular,* $(M \coprod N)^\vee = M^\vee \coprod N^\vee$ *for* $M, N \in \operatorname{Ref} R$.
(2) $M^{\vee\vee} = M$ *for every* $M \in \operatorname{Ref} R$.
(3) $M^\vee$ *is a direct summand of* M^* *for every* $M \in \operatorname{Ref} R$.

Moreover, any assignment between right and left R-modules satisfying (1) - (3) *sends a pure-reflexive module M to $M^\vee$.*

REMARK 4.15. An assignment between right and left R-modules which satisfies (2) and (3) is only possible between pure-injective modules. This follows from Lemma 4.5

We begin the proof of this theorem with an observation which is of interest in itself.

PROPOSITION 4.16. *Every pure-reflexive module M is purely opposed to $M^\vee$.*

PROOF. The assertion follows from Proposition 4.12 if M is indecomposable. Now let $M = \widehat{\coprod}_{i\in I} M_i$ with $M_i \in \operatorname{Ind} R$ for all i. Clearly, $\coprod_{i\in I} M_i$ and $\coprod_{i\in I} M_i^\vee$ are purely opposed, and from this follows that M and $M^\vee$ are purely opposed since every module is purely equivalent to its pure-injective envelope by Proposition 4.6. □

PROOF OF THEOREM 4.14. (1) follows from Lemma 4.13, and (2) is a consequence of Proposition 4.12 in combination with (1). It remains to check (3). Observe first that M^* is purely opposed to M by Lemma 4.4 and therefore purely equivalent to $M^\vee$ by Proposition 4.16. If M is indecomposable, then the construction of $M^\vee$ given in Proposition 4.12 shows that $M^\vee$ is a direct summand of M^* since both modules are purely-equivalent. Now assume that $M = \widehat{\coprod}_{i\in I} M_i$ with $M_i \in \operatorname{Ind} R$ for all i. The pure monomorphism $\coprod_{i\in I} M_i \to M$ induces a split epimorphism $M^* \to (\coprod_{i\in I} M_i)^*$. We obtain a pure monomorphism

$$\coprod_{i\in I} M_i^\vee \longrightarrow \prod_{i\in I} M_i^\vee \longrightarrow \prod_{i\in I} M_i^* \xrightarrow{\sim} (\coprod_{i\in I} M_i)^* \longrightarrow M^*$$

and therefore $M^\vee$ is a direct summand of M^* since M^* is pure-injective.

We prove now the last statement. Suppose there are classes $\mathcal{X}$ in $\operatorname{Mod} R$ and $\mathcal{Y}$ in $\operatorname{Mod} R^{\mathrm{op}}$ which are closed under direct summands, and suppose there are maps $M \mapsto M'$ between $\mathcal{X}$ and $\mathcal{Y}$ which satisfy (1) - (3). It follows from (2) and (3) with Lemma 4.4 that M and M' are purely opposed, and Lemma 4.5 implies that every module in $\mathcal{X}$ and $\mathcal{Y}$ is pure-injective. Furthermore, (1) and (2) imply that M' is indecomposable if M is indecomposable. Therefore $M' = M^\vee$ for every indecomposable $M \in \operatorname{Ref} R$ which lies in $\mathcal{X}$ by Proposition 4.12. Now let $M = \widehat{\coprod}_{i\in I} M_i$ be an arbitrary pure-reflexive R-module which lies in $\mathcal{X}$. It follows that $M_i \in \mathcal{X}$ for all i since $\mathcal{X}$ is closed under direct summands, and therefore

$$M' = (\widehat{\coprod_{i\in I}} M_i)' = \widehat{\coprod_{i\in I}} M_i' = \widehat{\coprod_{i\in I}} M_i^\vee = M^\vee.$$

This completes the proof of the theorem. □

We observe that the map $M \mapsto M^\vee$ is compatible with the Ziegler topology.

PROPOSITION 4.17. *Let* $\mathbf{U} \subseteq \operatorname{Ind} R$ *be a set of pure-reflexive modules. If* $M \in \operatorname{Ind} R$ *is pure-reflexive, then* M *belongs to the Ziegler closure* $\overline{\mathbf{U}}$ *if and only if* $M^\vee \in \overline{\mathbf{U}^\vee}$.

PROOF. Let $\mathcal{S}_{\mathbf{U}}$ be the Serre subcategory of all objects F in $C(R)$ satisfying $\operatorname{Hom}(F, T_N) = 0$ for every $N \in \mathbf{U}$. Then $M \in \overline{\mathbf{U}}$ if and only if $\operatorname{Hom}(\mathcal{S}_{\mathbf{U}}, T_M) = 0$. We have $(\mathcal{S}_{\mathbf{U}})^\vee = \mathcal{S}_{\mathbf{U}^\vee}$ and therefore $M \in \overline{\mathbf{U}}$ if and only if $M^\vee \in \overline{\mathbf{U}^\vee}$ since M and $M^\vee$ are purely opposed. □

CHAPTER 5

Ideals in the category of finitely presented modules

In this chapter we introduce a new class of ideals in the category $\operatorname{mod} R$ of finitely presented R-modules. These ideals are closely related to the definable subcategories of $\operatorname{Mod} R$.

5.1. Fp-idempotent ideals

Throughout this section R denotes an artin algebra over some fixed commutative artinian ring k. Our aim in this section is to establish an inclusion preserving bijection between certain ideals in $\operatorname{mod} R$ and the definable subcategories of $\operatorname{Mod} R$.

We begin with some definitions and refer to the appendix for some of the basic facts about ideals in additive categories. Recall that an *ideal* $\mathfrak{I}$ in $\operatorname{mod} R$ consists of subgroups $\mathfrak{I}(X,Y)$ in $\operatorname{Hom}_R(X,Y)$ for every pair X,Y in $\operatorname{mod} R$ such that for all maps $\alpha\colon X' \to X$ and $\beta\colon Y \to Y'$ in $\operatorname{mod} R$ the composition $\beta\circ\phi\circ\alpha$ belongs to $\mathfrak{I}(X',Y')$ for every $\phi \in \mathfrak{I}(X,Y)$. Note that an ideal $\mathfrak{I}$ in $\operatorname{mod} R$ is idempotent if and only if the class of functors in $(\operatorname{mod} R, \mathrm{Ab})$ vanishing on $\mathfrak{I}$ is closed under extensions. Therefore we call $\mathfrak{I}$ *fp-idempotent* if the class of finitely presented functors in $(\operatorname{mod} R, \mathrm{Ab})$ vanishing on $\mathfrak{I}$ is closed under extensions. Observe that any idempotent ideal is fp-idempotent; however the converse is usually not true. We introduce some further terminology. Given a functor $F\colon \operatorname{mod} R \to \mathrm{Ab}$, we denote by $\operatorname{ann} F$ the ideal of maps ϕ in $\operatorname{mod} R$ satisfying $F(\phi) = 0$. Given a collection Φ of maps and an ideal $\mathfrak{I}$ in $\operatorname{mod} R$ we define

$$\operatorname{ann}\Phi = \bigcap_{\phi\in\Phi} \operatorname{ann}\operatorname{Coker} H_\phi \quad \text{and} \quad \operatorname{ann}^{-1}\mathfrak{I} = \{\phi \in \operatorname{mod} R \mid \mathfrak{I} \subseteq \operatorname{ann}\operatorname{Coker} H_\phi\}.$$

Note that a map $\phi\colon X \to Y$ belongs to $\operatorname{ann}^{-1}\mathfrak{I}$ if and only if every $\psi\colon X \to Z$ in $\mathfrak{I}$ factors through ϕ. Given an ideal $\mathfrak{I}$ in $\operatorname{mod} R$ the collection $\operatorname{ann}^{-1}\mathfrak{I}$ can be used to characterize the fact that $\mathfrak{I}$ is fp-idempotent.

LEMMA 5.1. *$\mathfrak{I}$ is fp-idempotent if and only if $\phi', \phi'' \in \operatorname{ann}^{-1}\mathfrak{I}$ implies $\left[\begin{smallmatrix}\phi' & 0\\ \psi & \phi''\end{smallmatrix}\right] \in \operatorname{ann}^{-1}\mathfrak{I}$ for all $\psi \in \operatorname{mod} R$.*

PROOF. See Lemma C.10. □

Fp-idempotent ideals are usually not idempotent. However, we will show that each fp-idempotent ideal $\mathfrak{I}$ is of the form $\mathfrak{I} = \mathfrak{J} \cap \operatorname{mod} R$ for some idempotent ideal $\mathfrak{J}$ of $\operatorname{Mod} R$. We now state the main result of this chapter. To this end denote for any class $\mathcal{C}$ of R-modules by $[\mathcal{C}]$ the ideal of maps in $\operatorname{mod} R$ which factor through some finite coproduct of modules in $\mathcal{C}$.

THEOREM 5.2. *Let R be an artin algebra. The assignments*

$$\mathcal{X} \mapsto [\mathcal{X}] \quad \textit{and} \quad \mathfrak{I} \mapsto (\operatorname{Mod} R)_{\operatorname{ann}^{-1}\mathfrak{I}}$$

give mutually inverse and inclusion preserving bijections between the definable subcategories of $\operatorname{Mod} R$ *and the fp-idempotent ideals in* $\operatorname{mod} R$.

The proof of this result is given in two steps. In this section we discuss the basic properties of fp-idempotent ideals using finitely presented functors $\operatorname{mod} R \to \mathrm{Ab}$. We derive from these properties that the assignment $\mathfrak{I} \mapsto (\operatorname{Mod} R)_{\operatorname{ann}^{-1}\mathfrak{I}}$ induces a bijection between the fp-idempotent ideals and the definable subcategories. In the following section we discuss the connection between fp-idempotent ideals in $\operatorname{mod} R$ and idempotent ideals in $\operatorname{Mod} R$.

We begin with some notation. Denote by

$$\operatorname{Mod} k \longrightarrow \operatorname{Mod} k, \quad M \mapsto M^* = \operatorname{Hom}_k(M, E)$$

the functor given by an injective envelope E of $k/\operatorname{rad} k$. It is well-known that this induces a duality $\operatorname{mod} k \to \operatorname{mod} k$ and therefore also a duality $\operatorname{mod} R \to \operatorname{mod} R^{\mathrm{op}}$. We denote by $D_k(R) = (\operatorname{mod} R^{\mathrm{op}}, \operatorname{mod} k)$ the category of k-linear functors $\operatorname{mod} R^{\mathrm{op}} \to \operatorname{mod} k$. Given a functor $F \in D(R)$ we define $F^*\colon \operatorname{Mod} R \to \mathrm{Ab}$ by $F^*(X) = F(X^*)^*$. The same formula defines for every $F \in D(R^{\mathrm{op}})$ a functor $F^*\colon \operatorname{Mod} R^{\mathrm{op}} \to \mathrm{Ab}$.

LEMMA 5.3. *The following holds:*

(1) *The assignment* $F \mapsto F^*$ *induces mutually inverse equivalences between* $D_k(R)^{\mathrm{op}}$ *and* $D_k(R^{\mathrm{op}})$.
(2) *There is a functorial isomorphism* $F^*(X) \simeq F^\vee(X)$ *for every* $F \in D(R)$ *and* $X \in \operatorname{mod} R$.

PROOF. (1) $F^{**} \simeq F$ for every $F \in D_k(R)$ since $X^{**} \simeq X$ for all X in $\operatorname{mod} k$ and $\operatorname{mod} R$.

(2) The assignments $F \mapsto F^*|_{\operatorname{mod} R}$ and $F \mapsto F^\vee|_{\operatorname{mod} R}$ induce two functors $D(R) \longrightarrow D(R^{\mathrm{op}})$. Both are exact and preserve coproducts, i.e.

$$\Big(\coprod_{i\in I} F_i\Big)^* \simeq \prod_{i\in I} F_i^* \quad \text{and} \quad \Big(\coprod_{i\in I} F_i\Big)^\vee \simeq \prod_{i\in I} F_i^\vee$$

for every family $(F_i)_{i\in I}$ in $D(R)$. For $F = \operatorname{Hom}_{R^{\mathrm{op}}}(Y, -)$ with $Y \in \operatorname{mod} R^{\mathrm{op}}$ we have

$$F^*(X) \simeq X \otimes_R Y \simeq F^\vee(X).$$

It follows that $F^* \simeq F^\vee$ for arbitrary $F \in D(R)$ since F is a colomit of representable functors, i.e. there is an exact sequence

$$\coprod_{j\in J} \operatorname{Hom}_{R^{\mathrm{op}}}(Y_j, -) \longrightarrow \coprod_{i\in I} \operatorname{Hom}_{R^{\mathrm{op}}}(X_i, -) \longrightarrow F \longrightarrow 0$$

in $D(R)$ with $X_i, Y_j \in \operatorname{mod} R^{\mathrm{op}}$ for all i and j. □

Now let $\mathcal{S}$ be a Serre subcategory of $C(R^{\mathrm{op}})$. Let $\mathfrak{I} = \operatorname{ann} \mathcal{S}$ and denote by $\mathcal{T} = \varinjlim \mathcal{S}$ the corresponding localizing subcategory of $D(R^{\mathrm{op}})$. Given F in $D(R^{\mathrm{op}})$ we define

$$tF = \sum_{G\in\mathcal{S}} \operatorname{Im}(G \to F) \quad \text{and} \quad rF = \bigcap_{G\in\mathcal{S}} \operatorname{Ker}(F \to G).$$

Analogously, we consider the Serre subcategory $\mathcal{S}' = \mathcal{S}^*$ of $C(R)$ together with

$$t'F = \sum_{G\in\mathcal{S}'} \operatorname{Im}(G \to F) \quad \text{and} \quad r'F = \bigcap_{G\in\mathcal{S}'} \operatorname{Ker}(F \to G)$$

for every F in $D(R)$.

LEMMA 5.4. *The following holds:*

(1) $F \mapsto tF$ *assigns to each* F *in* $D(R^{\mathrm{op}})$ *its unique maximal subobject contained in* $\mathcal{T}$. *Therefore* $t\colon D(R^{\mathrm{op}}) \to \mathcal{T}$ *is the right adjoint of the inclusion* $\mathcal{T} \to D(R^{\mathrm{op}})$.

(2) $tF = (F^*/r'F^*)^*$ *and* $rF = (F^*/t'F^*)^*$ *for* $F \in D_k(R^{\mathrm{op}})$.

(3) $\mathfrak{I}(X,Y) = r\,\mathrm{Hom}_R(X,-)(Y)$ *for all* X, Y *in* $\mathrm{mod}\,R$.

PROOF. (1) Follows from the fact that a finitely generated subobject $F' \subseteq F$ belongs to $\mathcal{T}$ if and only if there is a morphism $\phi\colon G \to F$ with $\mathrm{Im}\,\phi = F'$ for some $G \in \mathcal{S}$.

(2) Use the fact that under the duality $D_k(R^{\mathrm{op}}) \to D_k(R)$ the morphisms $G \to F$ with $G \in \mathcal{S}$ correspond to the morphisms $F^* \to H$ with $H \in \mathcal{S}'$.

(3) By definition, a map $\phi\colon X \to Y$ belongs to $\mathrm{ann}\,\mathcal{S}$ if and only if $\phi \in \mathrm{Ker}\,\pi_Y$ for every $\pi\colon \mathrm{Hom}(X,-) \to F$ with $F \in \mathcal{S}$. Thus $\phi \in \mathfrak{I}$ if and only if $\phi \in r\,\mathrm{Hom}_R(X,-)(Y)$. □

LEMMA 5.5. *The following are equivalent for* F *in* $C(R^{\mathrm{op}})$:

(1) $F \in \mathcal{S}$.

(2) $F(\phi) = 0$ *for all* $\phi \in \mathfrak{I}$.

(3) $rF = 0$.

PROOF. (1) $\Rightarrow$ (2) Clear.

(2) $\Rightarrow$ (3) Applying the preceding lemma we have that $D_k(R^{\mathrm{op}}) \to D_k(R^{\mathrm{op}})$, $G \mapsto G/rG$, is right exact since $G/rG = (t'G^*)^*$ and t' is left exact. We obtain therefore the following commutative diagram with exact rows:

$$\begin{array}{ccccccc}
\mathrm{Hom}_R(Y,-) & \xrightarrow{\mathrm{Hom}(\psi,-)} & \mathrm{Hom}_R(X,-) & \xrightarrow{\pi} & F & \longrightarrow & 0 \\
\downarrow & & \downarrow & & \downarrow & & \\
\mathrm{Hom}_R(Y,-)/\mathfrak{I}(Y,-) & \longrightarrow & \mathrm{Hom}_R(X,-)/\mathfrak{I}(X,-) & \longrightarrow & F/rF & \longrightarrow & 0
\end{array}$$

Now assume $rF \neq 0$ and choose $\alpha\colon X \to Z$ with $0 \neq \pi_Z(\alpha) \in rF(Z)$. Using the commutativity of the diagram we find $\beta\colon Y \to Z$ with $\phi = \alpha - \beta \circ \psi \in \mathfrak{I}(X,Z)$, but $F(\phi) \neq 0$ since $F(\alpha) \neq 0$.

(3) $\Rightarrow$ (1) If $rF = 0$, then $t'F^* = F^*$ by the preceding lemma. Thus $F^* \in \mathcal{S}'$ and therefore $F \in \mathcal{S}$. □

The preceding lemma can be reformulated as follows.

LEMMA 5.6. *We have* $\Phi = \mathrm{ann}^{-1}(\mathrm{ann}\,\Phi)$ *for every saturated collection* Φ *of maps in* $\mathrm{mod}\,R$.

PROOF. Recall from Lemma 2.8 that the collection Φ is saturated if and only if $\mathcal{S} = \{\mathrm{Coker}\,H_\phi \mid \phi \in \Phi\}$ is a Serre subcategory of $C(R^{\mathrm{op}})$. By definition, we have $\Phi \subseteq \mathrm{ann}^{-1}(\mathrm{ann}\,\Phi)$. Conversely, if $\mathrm{ann}\,\Phi \subseteq \mathrm{Coker}\,H_\phi$ for some map ϕ, then $\mathrm{Coker}\,H_\phi \in \mathcal{S}$ by the preceding lemma, and therefore $\phi \in \Phi$. □

LEMMA 5.7. *If* $\mathfrak{I}$ *is an ideal in* $\mathrm{mod}\,R$, *then* $\mathfrak{I} = \mathrm{ann}(\mathrm{ann}^{-1}\mathfrak{I})$.

PROOF. Clearly, $\mathfrak{I} \subseteq \mathrm{ann}(\mathrm{ann}^{-1}\mathfrak{I})$. Now let $\phi\colon X \to Y$ be a map not in $\mathfrak{I}$. We need to find $F \in C(R^{\mathrm{op}})$ with $F(\mathfrak{I}) = 0$ and $F(\phi) \neq 0$. Take $G = \mathrm{Hom}(X,-)/\mathfrak{I}(X,-)$. Then $G^*(\phi^*) \neq 0$. This gives a map $\pi\colon \mathrm{Hom}(Y^*,-) \to G^*$ with $\pi \circ \mathrm{Hom}(\phi^*,-) \neq 0$. Take $F = (\mathrm{Im}\,\pi)^*$. Then $F(\phi) \neq 0$ but $F(\mathfrak{I}) = 0$ since F is a quotient of G and $G(\mathfrak{I}) = 0$. □

PROPOSITION 5.8. *Let R be an artin algebra. The assignments $\Phi \mapsto \operatorname{ann}\Phi$ and $\mathfrak{I} \mapsto \operatorname{ann}^{-1}\mathfrak{I}$ induce mutually inverse bijections between the set of saturated collections of maps in* $\operatorname{mod} R$ *and the set of fp-idempotent ideals in* $\operatorname{mod} R$.

PROOF. Combine Lemma 5.6 and Lemma 5.7. □

The main result of this section is now an immediate consequence.

COROLLARY 5.9. *The assignment $\mathfrak{I} \mapsto (\operatorname{Mod} R)_{\operatorname{ann}^{-1}\mathfrak{I}}$ induces an inclusion preserving bijection between the fp-idempotent ideals in* $\operatorname{mod} R$ *and the definable subcategories of* $\operatorname{Mod} R$.

PROOF. Combine the bijection $\mathfrak{I} \mapsto \operatorname{ann}^{-1}\mathfrak{I}$ from the preceding proposition with the bijection $\Phi \mapsto (\operatorname{Mod} R)_\Phi$ from Theorem 2.9. □

COROLLARY 5.10. *The assignments*

$$\mathcal{S} \mapsto \bigcap_{F \in \mathcal{S}} \operatorname{ann} F \quad \textit{and} \quad \mathfrak{I} \mapsto \{F \in \mathcal{C}(R^{\mathrm{op}}) \mid F(\mathfrak{I}) = 0\}$$

induce mutually inverse bijections between the set of Serre subcategories of $\mathcal{C}(R^{\mathrm{op}})$ and the set of fp-idempotent ideals in $\operatorname{mod} R$.

PROOF. This is a consequence of Proposition 5.8. □

COROLLARY 5.11. *Let $\mathfrak{I}$ be an ideal in* $\operatorname{mod} R$. *Then the set of fp-idempotent ideals contained in $\mathfrak{I}$ has a unique maximal element $\mathfrak{J}$. It satisfies $\bigcap_{n \in \mathbb{N}} \mathfrak{I}^n \subseteq \mathfrak{J} \subseteq \mathfrak{I}$.*

PROOF. Let $\mathcal{S}$ be the Serre subcategory of $\mathcal{C}(R^{\mathrm{op}})$ which is generated by all functors vanishing on $\mathfrak{I}$. Then $\mathfrak{J} = \operatorname{ann}\mathcal{S}$. The functors in $\mathcal{C}(R^{\mathrm{op}})$ vanishing on $\mathfrak{I}^n$ for some $n \in \mathbb{N}$ form a Serre subcategory $\mathcal{T}$ by Lemma C.8 which contains $\mathcal{S}$. Therefore

$$\bigcap_{n \in \mathbb{N}} \mathfrak{I}^n \subseteq \operatorname{ann}\mathcal{T} \subseteq \operatorname{ann}\mathcal{S} = \mathfrak{J}.$$

□

EXAMPLE 5.12. Let $\mathfrak{I} = \operatorname{rad}(\operatorname{mod} R)$ be the Jacobson radical of $\operatorname{mod} R$, i.e. $\mathfrak{I}$ is the intersection of all maximal ideals in $\operatorname{mod} R$. The unique maximal fp-idempotent ideal contained in $\mathfrak{I}$ is $\mathfrak{I}^\omega = \bigcap_{n \in \mathbb{N}} \mathfrak{I}^n$, see Theorem 8.12.

5.2. Ideals generated by idempotents

Throughout this section R denotes an artin algebra. We continue our discussion of fp-idempotent ideals in $\operatorname{mod} R$. These ideals are usually not idempotent. However, we will show that each fp-idempotent ideal is of the form $\mathfrak{I} \cap \operatorname{mod} R$ for some idempotent ideal $\mathfrak{I}$ of $\operatorname{Mod} R$. In fact, we shall find idempotent maps in $\operatorname{Mod} R$ which generate the ideal $\mathfrak{I}$.

We begin with some notation. Let $\mathcal{C}$ be any class of R-modules. We denote by $[\mathcal{C}]$ the ideal of maps in $\operatorname{mod} R$ which factor through some finite coproduct of modules in $\mathcal{C}$. The full subcategory of direct summands of arbitrary products of modules in $\mathcal{C}$ is denoted by $\operatorname{Prod}\mathcal{C}$.

LEMMA 5.13. *Let $\mathcal{S}$ be a collection of coherent functors* $\operatorname{Mod} R \to \operatorname{Ab}$ *and suppose that $\{F|_{\operatorname{mod} R} \mid F \in \mathcal{S}\}$ is a Serre subcategory of $\mathcal{C}(R^{\mathrm{op}})$. Then there exists for every R-module M, up to isomorphism, a unique map $\phi\colon M \to N$ having the following properties:*

(1) *N is pure-injective and $F(N) = 0$ for all F in $\mathcal{S}$.*
(2) *A map $\phi' \colon M \to N'$ into a pure-injective module N' factors through ϕ if and only if $F(\phi') = 0$ for all F in $\mathcal{S}$.*
(3) *ϕ is left minimal.*

PROOF. We have $\mathcal{S} = \mathcal{T}^{\vee}$ for some Serre subcategory $\mathcal{T}$ of $C(R)$ by Lemma 1.6. Let $\varinjlim \mathcal{T}$ be the localizing subcategory of $D(R)$ generated by $\mathcal{T}$ and denote by tT_M the maximal subobject of T_M which belongs to $\varinjlim \mathcal{T}$. Taking an injective envelope $T_M/tT_M \to T_N$ gives a map $\phi \colon M \to N$ which has the desired properties. To see this choose $F \in \mathcal{S}$. We have $F = G^{\vee}$ for some $G \in \mathcal{T}$ and therefore $F(L) = \operatorname{Hom}(G, T_L)$ for every R-module L. Thus $F(N) = 0$. Now let $\phi' \colon M \to N'$ be any map with N' being pure-injective. Clearly, $F(\phi') = 0$ if ϕ' factors through ϕ. Conversely, $F(\phi') = 0$ implies $tT_M \subseteq K$ for $K = \operatorname{Ker} T_{\phi'}$. The inclusion $tT_M \to K$ induces a map $TM/tT_M \to T_M/K$ and the composition with $T_M/K \to T_{N'}$ extends to a map $T_N \to T_{N'}$ since N' is pure-injective. Thus ϕ' factors through ϕ. Finally, ϕ' is left minimal since the injective envelope $T_M/tT_M \to T_N$ is left minimal. □

We can now prove that every fp-idempotent ideal is generated by idempotent maps in $\operatorname{Mod} R$. In fact, we show that $[\mathcal{X}]$ is the fp-idempotent ideal corresponding to a definable subcategory $\mathcal{X}$. This is precisely the contents of Theorem 5.2.

PROOF OF THEOREM 5.2. We use the bijective correspondence between fp-idempotent ideals and definable subcategories from Corollary 5.9. Fix an fp-idempotent ideal $\mathfrak{I}$ in $\operatorname{mod} R$ and let $\mathcal{X} = (\operatorname{Mod} R)_{\operatorname{ann}^{-1}\mathfrak{I}}$ be the corresponding definable subcategory of $\operatorname{Mod} R$. We need to show that $\mathfrak{I} = [\mathcal{X}]$. Let $\Phi = \operatorname{ann}^{-1}\mathfrak{I}$ and let $\mathcal{S} = \{\operatorname{Coker} H_{\phi} \mid \phi \in \Phi\}$. Note that an R-module M belongs to $\mathcal{X}$ if and only if $F(M) = 0$ for all F in $\mathcal{S}$. Now fix a map ϕ between finitely presented R-modules. We have $\phi \in \mathfrak{I}$ if and only if $F(\phi) = 0$ for all F in $\mathcal{S}$ by Proposition 5.8. It follows from Lemma 5.13 that $F(\phi) = 0$ for all F in $\mathcal{S}$ if and only if ϕ factors through a (pure-injective) module in $\mathcal{X}$. Therefore $\mathfrak{I} = [\mathcal{X}]$. □

We present now various consequences of Theorem 5.2.

COROLLARY 5.14. *The assignment $\mathbf{U} \mapsto [\operatorname{Prod} \mathbf{U}]$ induces an inclusion preserving bijection between the Ziegler-closed subsets of $\operatorname{Ind} R$ and the fp-idempotent ideals in $\operatorname{mod} R$.*

PROOF. The argument given in the proof of Theorem 5.2 shows that $[\mathcal{X}] = [\operatorname{Prod} \mathbf{U}]$ since the pure-injective modules in $\mathcal{X}$ are precisely the direct summands of modules in $\operatorname{Prod} \mathbf{U}$. The assignment $\mathbf{U} \mapsto [\operatorname{Prod} \mathbf{U}]$ is therefore the composition of the bijection between Ziegler-closed subsets and definable subcategories with the bijection between definable subcategories and fp-idempotent ideals. □

COROLLARY 5.15. *Let $\mathcal{X}$ be a definable subcategory of $\operatorname{Mod} R$ which is generated by a Σ-pure-injective R-module, and let $\mathbf{U} = \mathcal{X} \cap \operatorname{Ind} R$. Then $[\mathcal{X}] = [\mathbf{U}] = [\operatorname{Prod} \mathbf{U}]$.*

PROOF. The assumption on $\mathcal{X}$ implies that every module in $\mathcal{X}$ is a coproduct of modules in $\mathbf{U}$. This follows from Corollary 2.7; see also Lemma 6.5. The assertion now follows since every map $X \to \coprod_{i \in I} M_i$ with $X \in \operatorname{mod} R$ factors through $\coprod_{i \in J} M_i$ for some finite subset $J \subseteq I$. □

COROLLARY 5.16. *Let $\mathbf{U} \subseteq \operatorname{Ind} R$ be a set of finitely presented modules. Then $[\mathbf{U}]$ is the fp-idempotent ideal corresponding to the Ziegler closure $\overline{\mathbf{U}}$.*

PROOF. $[\mathbf{U}]$ is fp-idempotent and $\Phi_{\mathbf{U}} = \operatorname{ann}^{-1}[\mathbf{U}]$. □

COROLLARY 5.17. *If R is of finite representation type, then every fp-idempotent ideal of* $\operatorname{mod} R$ *is idempotent.*

Our discussion of generic modules over algebras of infinite representation type will provide examples of fp-idempotent ideals which are nilpotent and therefore not idempotent; see Theorem 6.28.

CHAPTER 6

Endofinite modules

Let M be an R-module. Then M can be regarded as an $\operatorname{End}_R(M)^{\mathrm{op}}$-module, and its length as an $\operatorname{End}_R(M)^{\mathrm{op}}$-module is called the *endolength* of M which is denoted by $\ell_{\mathrm{end}}(M)$. Following Crawley-Boevey, the module M is said to be *endofinite* if M is of finite endolength [**17**]. In this chapter we present various characterizations of endofiniteness and study the endofinite modules in the Ziegler spectrum. Our discussion includes two other classes of pure-injective modules. First we study Σ-pure-injective modules, and then we introduce a class of modules which we call product-complete since a module M is product-complete if $\operatorname{Add} M$ is closed under taking products. Every endofinite module is product-complete, and every product-complete module is Σ-pure-injective. The basic tool for our analysis of these finiteness conditions is the endocategory of a module.

6.1. The endocategory of a module

We fix an R-module M and let $S = \operatorname{End}_R(M)^{\mathrm{op}}$. We assign to M an abelian category $\mathcal{E}_M$ which is a subcategory of $\operatorname{Mod} S$. This endocategory of M was first introduced in [**47**] and the material in this section is taken from this paper. Consider the exact functor

$$E_M \colon C(R) \longrightarrow \operatorname{Mod} S, \quad F \mapsto \operatorname{Hom}(F, T_M)$$

where S acts on $E_M(F)$ via the isomorphism $S \simeq \operatorname{End}(T_M)^{\mathrm{op}}$. We denote by $\mathcal{S}_M$ the kernel of E_M which is a Serre subcategory of $C(R)$. The *endocategory* $\mathcal{E}_M$ of M is, by definition, the image of the induced functor $C(R)/\mathcal{S}_M \to \operatorname{Mod} S$ and we shall tacitly assume that this subcategory of $\operatorname{Mod} S$ is closed under isomorphisms. We need a further definition. Recall that an additve subgroup of M is of *finite definition* if it arises as the kernel of a morphism $M \to M \otimes_R X$, $m \mapsto m \otimes x$, for some finitely presented R^{op}-module X, and some element $x \in X$. Note that a subgroup of finite definition is a S-submodule of M.

Lemma 6.1. *The endocategory $\mathcal{E}_M$ of M has the following properties:*

(1) *$\mathcal{E}_M$ is abelian and the inclusion $\mathcal{E}_M \to \operatorname{Mod} S$ is exact.*
(2) *E_M induces an equivalence $(C(R)/\mathcal{S}_M)^{\mathrm{op}} \to \mathcal{E}_M$.*
(3) *The S-module M belongs to $\mathcal{E}_M$ and the subobjects of M in $\mathcal{E}_M$ are precisely the subgroups of finite definition.*
(4) *Every object in $\mathcal{E}_M$ is a subquotient of a finite coproduct of copies of M.*

Proof. (1) and (2) follow immediately from the definition of $\mathcal{E}_M$.

(3) We observe first that M belongs to $\mathcal{E}_M$ since $M \simeq \operatorname{Hom}(\operatorname{Hom}_{R^{\mathrm{op}}}(R, -), T_M)$ by Yoneda's lemma. A subgroup $U \subseteq M$ belongs to $\mathcal{E}_M$ if and only if there is $F \in C(R)$ having a presentation $\operatorname{Hom}_{R^{\mathrm{op}}}(X, -) \to \operatorname{Hom}_{R^{\mathrm{op}}}(R, -) \to F \to 0$ such

that $U \simeq E_M(F)$. The latter condition is equivalent to $U \simeq \operatorname{Ker} T_\phi$ for some $\phi\colon R \to X$. Therefore $U \in \mathcal{E}_M$ if and only if U is a subgroup of finite definition.

(4) The assertion follows from the fact that every object in $C(R)$ is a subquotient of a finite coproduct of copies of $\operatorname{Hom}_{R^{\mathrm{op}}}(R, -)$. □

We shall also consider the quotient category $D(R)/\mathcal{T}_M$ where $\mathcal{T}_M = \varinjlim \mathcal{S}_M$ denotes the localizing subcategory of $D(R)$ which is generated by $\mathcal{S}_M$.

LEMMA 6.2. *The functor* $\operatorname{Mod} R \to D(R)/\mathcal{T}_M$, $X \mapsto T_X$, *induces an equivalence between the definable subcategory generated by* M *and the full subcategory of fp-injective objects in* $D(R)/\mathcal{T}_M$. *Moreover,*

$$\operatorname{fp}(D(R)/\mathcal{T}) \simeq C(R)/\mathcal{S}_M \simeq (\mathcal{E}_M)^{\mathrm{op}}.$$

PROOF. The first assertion is a direct consequence of Corollary 2.6. The second assertion follows from Lemma 6.1 in combination with Proposition A.5. □

The definition of the endocategory implies immediately that two purely equivalent modules M and N have equivalent endocategories, i.e. $\mathcal{E}_M \simeq \mathcal{E}_N$. Analogously, two purely opposed modules M and N have anti-equivalent endocategories, i.e. $\mathcal{E}_M \simeq (\mathcal{E}_N)^{\mathrm{op}}$. This has some interesting consequences. To formulate them we denote for an R-module M by $\operatorname{Latt}(M)$ the lattice of subgroups of finite definition. Also we define $\Delta(M) = \operatorname{End}_R(M)/\operatorname{rad}\operatorname{End}_R(M)$.

PROPOSITION 6.3. *Let* M *be an* R*-module.*

(1) *Let* M *and* $N \in \operatorname{Mod} R$ *be purely equivalent. Then* $\operatorname{Latt}(M) \simeq \operatorname{Latt}(N)$.
(2) *Let* M *and* $N \in \operatorname{Mod} R^{\mathrm{op}}$ *be purely opposed. Then* $\operatorname{Latt}(M) \simeq \operatorname{Latt}(N)^{\mathrm{op}}$.
(3) *If* M *is indecomposable pure-reflexive, then* $\Delta(M^\vee) \simeq \Delta(M)^{\mathrm{op}}$.

PROOF. (1) and (2) follow from (3) in Lemma 6.1 since the endocategories of M and N are (anti-)equivalent.

(3) It follows from the characterization in Lemma 4.10 that $\Delta(M)$ is isomorphic to the endomorphism ring of the unique simple object in $\mathcal{E}_M$. The assertion now follows since M and $M^\vee$ are purely opposed by Proposition 4.16. □

6.2. Σ-pure-injective modules

A module M is called *Σ-pure-injective* if every coproduct of copies of M is pure-injective. In this section we collect the basic properties of such modules. The following characterization is well-known.

PROPOSITION 6.4. *The following are equivalent for a module* M*:*

(1) M *is* Σ*-pure-injective.*
(2) M *has the descending chain condition on subgroups of finite definition.*
(3) *There exists a cardinal* κ *such that every product of copies of* M *is a pure submodule of a coproduct of modules having cardinality at most* κ.

PROOF. See [**37, 84**]. □

We shall work with the following characterization of a Σ-pure-injective module.

LEMMA 6.5. M *is* Σ*-pure-injective if and only if* $D(R)/\mathcal{T}_M$ *is locally noetherian.*

PROOF. We use the fact that the finitely presented objects in $D(R)/\mathcal{T}_M$ form a generating set of objects for $D(R)/\mathcal{T}_M$. Then it follows from the characterization in part (2) of the preceding proposition in combination with Lemma 6.1 and Lemma 6.2 that M is Σ-pure-injective if and only if $D(R)/\mathcal{T}_M$ has a generating set of noetherian objects. □

The definable subcategory which is generated by a Σ-pure-injective module has the following properties.

PROPOSITION 6.6. *Let M be a Σ-pure-injective module.*

(1) *The direct summands of products of copies of M form a definable subcategory of* $\operatorname{Mod} R$.
(2) *Every direct summand of a product of copies of M is a coproduct of indecomposable modules with local endomorphism ring.*

PROOF. We use the fact that $D(R)/\mathcal{T}_M$ is locally noetherian.

(1) Every finitely generated object in $D(R)/\mathcal{T}_M$ is finitely presented by Proposition A.11. It follows that T_M is an injective cogenerator for $D(R)/\mathcal{T}_M$ and therefore every fp-injective object in $D(R)/\mathcal{T}_M$ is a direct summand of a product of copies of T_M. The assertion now follows from Lemma 6.2.

(2) Use again Proposition A.11 and Lemma 6.2. □

6.3. Product-complete modules

In this section we study a class of pure-injective modules which strictly contains the endofinite modules. Given any R-module M we denote by $\operatorname{Add} M$ the full subcategory of R-modules which are direct summands of coproducts of copies of M. Note that $\operatorname{Add} M$ is the smallest full subcategory of $\operatorname{Mod} R$ which contains M and is closed under forming arbitrary coproducts and direct summands. Our main result is a characterization of the fact that $\operatorname{Add} M$ is a definable subcategory.

THEOREM 6.7. *The following are equivalent for an R-module M:*

(1) $\operatorname{Add} M$ *is closed under taking products.*
(2) *Every product of copies of M is a direct summand of a coproduct of copies of M.*
(3) *Every product of copies of M is a coproduct of (indecomposable) direct summands of M.*
(4) *M is Σ-pure-injective and the indecomposable direct summands of M form a Ziegler-closed subset of* $\operatorname{Ind} R$.
(5) *The coproducts of indecomposable direct summands of M form a definable subcategory.*
(6) $\operatorname{Add} M$ *is a definable subcategory.*

We call a module M *product-complete* if M satisfies the equivalent conditions of the preceding theorem.

PROOF. (1) $\Rightarrow$ (2) Clear.

(2) $\Rightarrow$ (3) M is Σ-pure-injective by Proposition 6.4. It follows from Proposition 6.6 in combination with the Krull-Remak-Schmidt-Azumaya Theorem that every direct summand of a coproduct of copies of M decomposes into indecomposable modules which are direct summands of M. Thus (2) implies (3).

(3) ⇒ (4) M is Σ-pure-injective by Proposition 6.4. It follows from (3) and Proposition 6.6 that every indecomposable module belonging to the definable subcategory generated by M is a direct summand of M. Thus the indecomposable direct summands of M form a Ziegler-closed subset of $\operatorname{Ind} R$.

(4) ⇒ (5) Use again Proposition 6.6.

(5) ⇒ (6) Clear.

(6) ⇒ (1) A definable subcategory is closed under taking products. □

The following consequence is a joint result with Saorín [**55**].

COROLLARY 6.8. *An R-module M is product-complete if and only if every R-module has a (minimal) left* $\operatorname{Add} M$*-approximation.*

PROOF. Combine the preceding theorem with Corollary 3.15. □

We give an example of a Σ-pure-injective module which is not product-complete.

EXAMPLE 6.9. Let $R = \mathbb{Z}$ and let $M = \mathbb{Z}_{p^\infty}$ be a Prüfer group. This module is Σ-pure-injective but not product-complete. However $M \coprod \mathbb{Q}$ is product-complete.

We list some basic properties of product-complete modules.

(1) The class of product-complete modules is closed under finite coproducts.

(2) Let I be any non-empty set. Then a module M is product-complete if and only if $M^{(I)}$ is product-complete.

(3) If M is product-complete, then M^I is product-complete for every set I.

(4) A module M is Σ-pure-injective if and only if there exists a non-empty set I such that M^I is product-complete. In fact, if M is Σ-pure-injective, then one chooses I big enough such that for any set J any indecomposable direct summand of M^J already occurs as a direct summand of M^I.

(5) If M is product complete, then every subgroup of finite definition of M is a finitely generated module over $\operatorname{End}_R(M)^{\text{op}}$. In fact, a Σ-pure-injective module M is product complete if and only if every subgroup of finite definition of M is a finitely generated module over $\operatorname{End}_R(M)^{\text{op}}$, see [**55**].

6.4. Endofinite modules

In this section we present a number of characterizations of endofiniteness and study the endofinite modules in the Ziegler spectrum. Note that every endofinite module is automatically Σ-pure-injective; this follows from Proposition 6.4. In fact, most properties of endofinite modules can be derived from properties of Σ-pure-injective modules. Our first characterization of an endofinite module is based on its endocategory.

PROPOSITION 6.10. *An R-module M is endofinite if and only if the endocategory of M is a length category, i.e. each object is of finite length. Moreover, the endolength of M is precisely the length of M in its endocategory.*

PROOF. M is endofinite if and only if the lattice of subgroups of finite definition has the ascending and the descending chain condition. Moreover, in this case the subgroups of finite definition are precisely the $\operatorname{End}_R(M)^{\text{op}}$-submodules of M [**18**, Proposition 4.1]. Using Lemma 6.1, the assertion follows. □

REMARK 6.11. If M is endofinite with $S = \operatorname{End}_R(M)^{\text{op}}$, then the endocategory of M is a full subcategory of $\operatorname{Mod} S$ consisting of finite length modules.

Given an R-module M, we keep the notation from the beginning of this chapter. We denote by $\mathcal{S}_M$ the Serre subcategory of all F in $C(R)$ such that $\operatorname{Hom}(F, T_M) = 0$, and $\mathcal{T}_M = \varinjlim \mathcal{S}_M$ denotes the corresponding localizing subcategory of $D(R)$. We continue with some technical lemmas which are needed for further characterizations of endofiniteness.

LEMMA 6.12. *The endolength of an R-module M equals the length of the object* $\operatorname{Hom}_{R^{\mathrm{op}}}(R, -)$ *in* $C(R)/\mathcal{S}_M$.

PROOF. Combine Proposition 6.10 with (2) in Lemma 6.1. □

LEMMA 6.13. *M is endofinite if and only if* $D(R)/\mathcal{T}_M$ *is locally finite.*

PROOF. It follows from Proposition 6.10 and Lemma 6.2 that M is endofinite if and only if every object in $\operatorname{fp}(D(R)/\mathcal{T}_M) \simeq (\mathcal{E}_M)^{\mathrm{op}}$ is of finite length. An equivalent condition is that $D(R)/\mathcal{T}_M$ has a generating set of finite length objects, i.e. $D(R)/\mathcal{T}_M$ is locally finite. □

The next result, with the exception of part (2), is due to Gruson [**36**] and Garavaglia [**31**].

THEOREM 6.14. *The following are equivalent for a non-zero R-module M:*

(1) *M is indecomposable and endofinite.*
(2) *The coproducts of copies of M form a definable subcategory.*
(3) *M is indecomposable and every product of copies of M is a coproduct of copies of M.*

PROOF. (1) ⇒ (2) The preceding lemma implies that $D(R)/\mathcal{T}_M$ is locally noetherian. Therefore every fp-injective object in $D(R)/\mathcal{T}_M$ is a coproduct of indecomposable injective objects by Proposition A.11. Moreover, every indecomposable injective object in $D(R)/\mathcal{T}_M$ is the injective envelope of a simple object since $D(R)/\mathcal{T}_M$ is locally finite. It follows that T_M is, up to isomorphism, the unique indecomposable injective object in $D(R)/\mathcal{T}_M$ since M is indecomposable and $\operatorname{Hom}(S, T_M) \neq 0$ for every simple object S in $D(R)/\mathcal{T}_M$. Therefore every fp-injective object in $D(R)/\mathcal{T}_M$ is a coproduct of copies of T_M, and (2) follows.

(2) ⇒ (3) Clear, since the definable subcategory generated by M is closed under taking products.

(3) ⇒ (1) The module M is Σ-pure-injective by Proposition 6.4, and therefore $D(R)/\mathcal{T}_M$ is locally noetherian by Lemma 6.5. By construction, T_M is an injective cogenerator for $D(R)/\mathcal{T}_M$, and it follows from (3) that T_M is, up to isomorphism, the only indecomposable injective object in $D(R)/\mathcal{T}_M$. Thus $D(R)/\mathcal{T}_M$ coincides with its localizing subcategory generated by the finite length objects. It follows that $D(R)/\mathcal{T}_M$ is locally finite, and therefore M is endofinite by Lemma 6.13. □

The following characterization is due to Crawley-Boevey [**18**].

PROPOSITION 6.15. *M is endofinite if and only if there are indecomposable endofinite modules* $M_1, \ldots, M_n$ *and sets* $I_1, \ldots, I_n$ *such that* $M \simeq \coprod_{i=1}^n M_i^{(I_i)}$.

PROOF. We observe first that for every coproduct $M = \coprod_{i=1}^n M_i$ of pairwise non-isomorphic modules with local endomorphism ring we have $\ell_{\mathrm{end}}(M) = \sum_{i=1}^n \ell_{\mathrm{end}}(M_i)$ since

$$\operatorname{End}_R(M)/\operatorname{rad}\operatorname{End}_R(M) \simeq \prod_{i=1}^n \operatorname{End}_R(M_i)/\operatorname{rad}\operatorname{End}_R(M_i).$$

Suppose now that M is endofinite. Then M is Σ-pure-injective and has therefore a decomposition $M = \coprod_{i\in I} M_i^{(I_i)}$ for some subset $\{M_i \mid i \in I\}$ of $\operatorname{Ind} R$ by Proposition 6.6. Our first observation then shows that I needs to be finite. Conversely, if $M \simeq \coprod_{i=1}^{n} M_i^{(I_i)}$, then $\ell_{\mathrm{end}}(M) \leq \sum_{i=1}^{n} \ell_{\mathrm{end}}(M_i)$, and therefore M is endofinite if each M_i is endofinite. □

The next theorem generalizes the result of Gruson and Garavaglia; it is taken from [**55**].

THEOREM 6.16. *A module M is endofinite if and only if every direct summand of M is product-complete.*

PROOF. If M is indecomposable, then the assertion is precisely Theorem 6.14. Now suppose that M is an arbitrary module. We use the preceding proposition. The fact that every endofinite module is product-complete is then a consequence of the basic properties of product-complete modules listed after Theorem 6.7, and the fact that every indecomposable endofinite module is product-complete. Conversely, let M be a module such that every direct summand is product-complete. It follows from Theorem 6.7 that the indecomposable direct summands of M form a Ziegler-closed subset $\mathbf{U}$ of $\operatorname{Ind} R$ having the property that every subset $\mathbf{V} \subseteq \mathbf{U}$ is also Ziegler-closed. However, $\operatorname{Ind} R$ is a quasi-compact space by Proposition 2.10, and therefore $\mathbf{U}$ needs to be finite. Thus M is endofinite since each $N \in \mathbf{U}$ is endofinite and M is a coproduct of modules in $\mathbf{U}$ by Theorem 6.7. This completes the proof. □

Our next aim is to give some characterizations of endofinite modules which are formulated in terms of the Ziegler spectrum.

PROPOSITION 6.17. *The following are equivalent for a module M in* $\operatorname{Ind} R$*:*

(1) *M is endofinite.*
(2) *$\{M\}$ is Ziegler-closed and M is pure-reflexive.*
(3) *$\{M\}$ is Ziegler-closed and every non-zero direct summand of a product of copies of M has an indecomposable direct summand.*

PROOF. (1) $\Rightarrow$ (2) The set $\{M\}$ is Ziegler-closed by Theorem 6.14. The endocategory $\mathcal{E}_M$ contains a simple object since $\mathcal{E}_M$ is a length category by Proposition 6.10, and it follows from Lemma 4.10 that M is pure-reflexive.

(2) $\Rightarrow$ (3) It follows from Lemma 4.10 that $D(R)/\mathcal{T}_M$ contains a simple object S since M is pure-reflexive. The fact that $\{M\}$ is Ziegler-closed implies that $\operatorname{Hom}(S, T_N) \neq 0$ for every non-zero module N belonging to the definable subcategory generated by M. Therefore the injective envelope of S in $D(R)/\mathcal{T}_M$ is isomorphic to a direct summand of T_N which is indecomposable. This implies (3).

(3) $\Rightarrow$ (1) Let T_N be the injective envelope of a finitely presented object $X \neq 0$ in $D(R)/\mathcal{T}_M$. Assuming (3) it follows that N is a direct summand of a product of copies of M and has therefore an indecomposable direct summand which needs to be isomorphic to a module in $\{M\}$. The category $D(R)/\mathcal{T}_M$ is locally coherent and has therefore a simple object S which is unique since $\operatorname{Hom}(S, T_M) \neq 0$. We may assume that S is a subobject of T_M and that T_M is a direct summand of T_N. By construction $X \cap T_M \neq 0$ and therefore S is a finitely presented object since it is a finitely generated subobject of X. The bijection between Ziegler-closed subsets of $\operatorname{Ind} R$ and Serre subcategories of $C(R)$ shows that $\operatorname{fp}(D(R)/\mathcal{T}_M) \simeq C(R)/\mathcal{S}_M$ coincides with the Serre subcategory generated by S. Therefore $\operatorname{fp}(D(R)/\mathcal{T}_M)$ is a length category and M is endofinite by Lemma 6.13. □

PROPOSITION 6.18. *An R-module M is endofinite if and only if there is a (finite and Ziegler-closed) subset $\mathbf{U} \subseteq \operatorname{Ind} R$ such that the induced Ziegler topology on $\mathbf{U}$ is discrete and every product of copies of M is a coproduct of modules in $\mathbf{U}$.*

PROOF. One direction is an immediate consequence of Theorem 6.7 and Theorem 6.16 if we take for $\mathbf{U}$ the indecomposable direct summands of M. Therefore suppose that there exists a subset $\mathbf{U} \subseteq \operatorname{Ind} R$ such that the induced Ziegler topology on $\mathbf{U}$ is discrete and every product of copies of M is a coproduct of modules in $\mathbf{U}$. It follows from Proposition 6.4 that M is Σ-pure-injective. We may assume that each $N \in \mathbf{U}$ belongs to the definable subcategory generated by M. This implies that $\mathbf{U}$ is Ziegler-closed by Proposition 6.6. Moreover, every direct summand N of M is Σ-pure-injective and the indecomposable direct summands of N form a Ziegler-closed set since $\mathbf{U}$ carries the discrete topology. Thus every direct summand of M is product-complete by Theorem 6.7, and therefore M is endofinite by Theorem 6.16. $\square$

Recall that we have constructed a duality map $\operatorname{Ref} R \to \operatorname{Ref} R^{\mathrm{op}}$, $M \mapsto M^\vee$, between the pure-reflexive modules over R and R^{op}, respectively.

COROLLARY 6.19. *The assignment $M \mapsto M^\vee$ induces for every $n \in \mathbb{N}$ a bijection between the isomorphism classes of endofinite R-modules of endolength n and the isomorphism classes of endofinite R^{op}-modules of endolength n. Moreover, if $M = \coprod_{i \in I} M_i$ is endofinite, then $M^\vee = \coprod_{i \in I} M_i^\vee$.*

Using so-called characters, this bijection between the endofinite right and left modules has been established by Crawley-Boevey [**18**]. Here, we use the bijection $M \mapsto M^\vee$ between $\operatorname{Ref} R$ and $\operatorname{Ref} R^{\mathrm{op}}$.

PROOF OF COROLLARY 6.19. Every endofinite module is a coproduct of indecomposable pure-reflexive modules; this follows from Proposition 6.15 and Theorem 6.17. Therefore every endofinite module is pure-reflexive. Any pure-reflexive module M is purely opposed to $M^\vee$ by Proposition 4.16, and therefore M and $M^\vee$ have the same endolength by Lemma 6.12. The assertion is now a consequence of Theorem 4.14. $\square$

6.5. Generic modules

The indecomposable endofinite modules which are not finitely presented are of particular interest; they are called *generic* [**17, 51**]. In this section we study some basic propreties of generic modules. We have already seen that endofiniteness is reflected by properties of the Ziegler spectrum, and we continue this dicussion with a result which is due to Herzog [**40**].

PROPOSITION 6.20. *For every $n \in \mathbb{N}$ the modules of endolength at most n form a Ziegler-closed subset of* $\operatorname{Ind} R$.

We need some preparations for the proof. Let $F = \operatorname{Hom}_{R^{\mathrm{op}}}(R, -)$ and denote a chain

$$0 = F_0 \subseteq F_1 \subseteq \ldots \subseteq F_n = F$$

of subobjects in $C(R)$ by $(F_i)_{0 \leq i \leq n}$.

LEMMA 6.21. *An R-module M is endofinite of endolength at most n if and only if for every chain $(F_i)_{0 \leq i \leq n+1}$ there exists $i \in \{0, \ldots, n\}$ such that $E_M(F_{i+1}/F_i) = 0$.*

PROOF. Let $S = \operatorname{End}_R(M)^{\mathrm{op}}$. We use the fact that all S-submodules of M are of finite definition if M is endofinite [**18**, Proposition 4.1]. It follows from Lemma 6.1 that $E_M\colon C(R) \to \operatorname{Mod} S$ induces a surjective morphism from the lattice of subobjects of F onto the lattice of subgroups of finite definition of M. Combining these facts the assertion follows. □

PROOF OF PROPOSITION 6.20. Fix a chain $c = (F_i)_{0 \le i \le n+1}$ and choose for every i a map ϕ_i in $\operatorname{mod} R$ with $F_{i+1}/F_i = \operatorname{Ker} T_{\phi_i}$. We have $M \in \mathbf{U}_{\phi_i}$ if and only if $E_M(F_{i+1}/F_i) = 0$ by Lemma 2.2, and $\mathbf{U}_c = \bigcup_{i=0}^{n} \mathbf{U}_{\phi_i}$ is Ziegler-closed. Therefore $\mathbf{U}_n = \bigcap_c \mathbf{U}_c$ is Ziegler-closed where c runs through all chains $c = (F_i)_{0 \le i \le n+1}$. It follows from the preceding lemma that $\mathbf{U}_n$ is precisely the set of all modules of endolength at most n, and this finishes the proof. □

The following result discusses endofinite modules over noetherian algebras.

PROPOSITION 6.22. *Let R be a noetherian algebra.*

(1) *If $M \in \operatorname{Ind} R$ is endofinite, then M is finitely presented if and only if $\{M\}$ is Ziegler-open.*
(2) *Let $\mathbf{U} \subseteq \operatorname{Ind} R$ be an infinite set of finitely presented endofinite modules. If there is $n \in \mathbb{N}$ such that the endolength of each module in $\mathbf{U}$ is bounded by n, then the Ziegler closure of $\mathbf{U}$ contains a generic module.*

PROOF. (1) See [**51**, Theorem 4.7].

(2) We follow an idea of Herzog [**40**]. The Ziegler closure $\overline{\mathbf{U}}$ is a quasi-compact space by Proposition 2.10, consisting of endofinite modules by Proposition 6.20. It follows that $\{M\}$ is not open for some $M \in \overline{\mathbf{U}}$, and this module cannot be finitely presented by (1). □

We have the following description of generic modules in terms of Serre subcategories of $C(R)$.

PROPOSITION 6.23. *The assignment*

$$M \mapsto \mathcal{S}_M = \{F \in C(R) \mid \operatorname{Hom}(F, T_M) = 0\}$$

induces a bijection between

(1) *the isomorphism classes of indecomposable endofinite R-modules, and*
(2) *the maximal Serre subcategories $\mathcal{S}$ of $C(R)$ such that $C(R)/\mathcal{S}$ contains a simple object.*

If R is a noetherian algebra, then M is generic if and only if $\mathcal{S}_M$ contains all finite length objects of $C(R)$.

PROOF. The first part of the assertion follows from the characterization of indecomposable endofinite modules in Theorem 6.14 if we compose the assignment $M \mapsto \mathcal{S}_M$ with the bijective correspondence between Serre subcategories of $C(R)$ and definable subcategories of $\operatorname{Mod} R$ from Corollary 2.3.

An indecomposable endofinite module M over a noetherian algebra is finitely presented if and only if $\operatorname{Hom}(S, T_M) \neq 0$ for some simple object S in $C(R)$; this follows from Proposition 7.1 in [**18**]. We conclude that M is generic if and only if $\mathcal{S}_M$ contains all finite length objects of $C(R)$. □

The endofinite modules control the representation type of a ring. In order to illustrate this fact we include a result which is a reformulation of a theorem of Auslander. In its present form the result is stated in [**61**] and [**83**].

THEOREM 6.24. *A ring R is of finite representation type, i.e. R is right artinian and has only finitely many isomorphism classes of finitely presented modules, if and only if every R-module is endofinite.*

PROOF. A ring R is of finite representation type if and only if $C(R)$ is a length category [**3**]. In Proposition 6.10 it was shown that a module M is endofinite if and only if its endocategory $\mathcal{E}_M$ is a length category. Moreover, $\mathcal{E}_M$ is by construction a quotient category of $C(R)^{\mathrm{op}}$ and is therefore a length category if $C(R)$ has this property. The assertion now follows since $\mathcal{E}_M \simeq C(R)^{\mathrm{op}}$ if one takes for M the product of all modules in $\operatorname{Ind} R$. □

The existence of generic modules over artin algebras is closely related to the second Brauer-Thrall conjecture, that an algebra of infinite representation type is of strongly unbounded representation type. In fact, Crawley-Boevey has shown that every algebra with a generic module is of strongly unbounded representation type [**18**]. A general existence result for generic modules can be formulated as follows. Recall from [**51**] that a family $\mathcal{T} = (M_i)_{i\in\mathbb{N}}$ of non-zero modules form a *generalized tube* if there are maps $M_i \to M_{i+1}$ and $M_{i+1} \to M_i$ for every $i \in \mathbb{N}$ which induce exact sequences $0 \to M_i \to M_{i-1} \coprod M_{i+1} \to M_i \to 0$ for every $i \in \mathbb{N}$ where $M_0 = 0$. We say that $\mathcal{T}$ *belongs to the radical* of $\operatorname{mod} R$ if all maps $M_i \to M_{i+1}$ and $M_{i+1} \to M_i$ belong to the Jacobson radical of $\operatorname{mod} R$. For example, the modules belonging to a homogeneous tube of the Auslander-Reiten quiver of R form a generalized tube in the radical of $\operatorname{mod} R$.

THEOREM 6.25. *Let R be an artin algebra and suppose there exists a generalized tube $\mathcal{T}$ in the radical of* $\operatorname{mod} R$. *Then there exists a generic R-module in the definable subcategory which is generated by $\mathcal{T}$.*

PROOF. See [**51**, Corollary 8.6]. □

For example, every finite dimensional algebra R over some algebraically closed field admits a generalized tube in the radical of $\operatorname{mod} R$, see [**14**]. We end this discussion of endofinite modules with another example.

EXAMPLE 6.26. Let R be a tame hereditary artin algebra and denote by $\mathcal{X}$ the full subcategory of torsion-free divisible R-modules M, i.e. $\operatorname{Hom}_R(X, M) = 0$ and $\operatorname{Ext}^1_R(X, M) = 0$ for all regular R-modules X, see [**70**]. This subcategory is of the form $\operatorname{Add} Q$ for some generic R-module Q, and therefore every R-module M has a minimal left $\mathcal{X}$-approximation $M \to Q_M$. The module Q_M is a coproduct of copies Q, and the rank of M in the sense of Ringel [**70**] is precisely the cardinal κ such that $Q_M = Q^{(\kappa)}$.

6.6. Ideals of finite length

Throughout this section R denotes an artin algebra. It has been shown in Theorem 5.2 that each definable subcategory of $\operatorname{Mod} R$ corresponds to an fp-idempotent ideal in $\operatorname{mod} R$. We wish to describe those ideals which correspond to a definable subcategory consisting of endofinite modules. To this end we define the length of an ideal in $\operatorname{mod} R$, and we establish a bijection between the fp-idempotent ideals of finite length and the finite subsets of $\operatorname{Ind} R$ consisting of endofinite modules.

Let $\mathfrak{I}$ be an ideal in $\operatorname{mod} R$. An *$\mathfrak{I}$-sequence of length n* is a sequence of maps

$$\begin{array}{ccccccccc} X_0 & \xrightarrow{\phi_1} & X_1 & \xrightarrow{\phi_2} & \cdots & \xrightarrow{\phi_{n-1}} & X_{n-1} & \xrightarrow{\phi_n} & X_n \\ \downarrow{\scriptstyle\psi_1} & & \downarrow{\scriptstyle\psi_2} & & & & \downarrow{\scriptstyle\psi_n} & & \\ Y_1 & & Y_2 & & & & Y_n & & \end{array}$$

in $\operatorname{mod} R$ such that each ψ_i belongs to $\mathfrak{I}$ and does not factor through ϕ_i. The *length* of $\mathfrak{I}$ is the maximal length of an $\mathfrak{I}$-sequence with $X_0 = R$ (or ∞ if such a natural number does not exist). The following lemma explains our definition.

LEMMA 6.27. *Let $\mathfrak{I}$ be an fp-idempotent ideal in* $\operatorname{mod} R$. *Then the length of $\mathfrak{I}$ equals the length of* $\operatorname{Hom}_R(R,-)$ *in $C(R^{\mathrm{op}})/\mathcal{S}$ where $\mathcal{S} = \{F \in C(R^{\mathrm{op}}) \mid F(\mathfrak{I}) = 0\}$.*

PROOF. Let $(\phi_i, \psi_i)\colon X_{i-1} \to X_i \coprod Y_i$, $1 \le i \le n$, be an $\mathfrak{I}$-sequence of length n. Let $F_0 = \operatorname{Hom}_R(X_0,-)$ and $F_i = \operatorname{Im}\operatorname{Hom}_R(\phi_i \circ \ldots \phi_1, -)$ for $1 \le i \le n$. The condition on each ψ_i implies that $(F_{i-1}/F_i)(\mathfrak{I}) \neq 0$ for all i and therefore $\operatorname{Hom}_R(X_0,-)$ has at least length n in $C(R^{\mathrm{op}})/\mathcal{S}$. Conversely, any chain $F_n \subseteq \ldots \subseteq F_0$ of subobjects in $C(R^{\mathrm{op}})$ with $F_{i-1}/F_i \notin \mathcal{S}$ for all i gives an $\mathfrak{I}$-sequence of length n. To this end choose epimorphisms $\operatorname{Hom}_R(X_i,-) \to F_i$. The inclusions $F_i \to F_{i-1}$ induce maps $\phi_i\colon X_{i-1} \to X_i$, and the assumption on each F_{i-1}/F_i implies the existence of maps $\psi_i\colon X_{i-1} \to Y_i$ in $\mathfrak{I}$ which do not factor through ϕ_i. □

THEOREM 6.28. *Let R be an artin algebra and suppose that M is an endofinite R-module of endolength n. Then $[M]$ is the fp-idempotent ideal corresponding to the definable subcategory of* $\operatorname{Mod} R$ *which is generated by M. Moreover, the following are equivalent:*

(1) *M has no finitely presented indecomposable direct summand.*
(2) *$[M]$ contains no invertible map.*
(3) *$[M]$ is nilpotent*
(4) *$[M]^{n+1} = 0$.*

PROOF. The module M is product-complete by Theorem 6.16, and therefore every module belonging to the definable subcategory $\mathcal{X}$ which is generated by M, is a direct summand of a coproduct of copies of M. Therefore $[\mathcal{X}] = [M]$ since every map $X \to M^{(I)}$ with X finitely presented factors through $M^{(J)}$ for some finite subset $J \subseteq I$. It follows from Theorem 5.2 that $[M]$ is the fp-idempotent ideal corresponding to $\mathcal{X}$. The equivalence (1) $\Leftrightarrow$ (2) is straightforward and (2) $\Leftrightarrow$ (3) $\Leftrightarrow$ (4) follows from the fact that $\operatorname{rad}^n \operatorname{End}_R(M) = 0$ by Nakayama's lemma. □

REMARK 6.29. Denote by $\mathfrak{rad} = \operatorname{rad}(\operatorname{mod} R)$ the Jacobson radical of $\operatorname{mod} R$ and let $\mathfrak{rad}^\omega = \bigcap_{n\in\mathbb{N}} \mathfrak{rad}^n$. If M is endofinite without finitely presented indecomposable direct summands, then $[M] \subseteq \mathfrak{rad}^\omega$. This follows from Corollary 8.13.

We are now in a position to formulate the precise relation between endofinite modules and fp-idempotent ideals of finite length.

COROLLARY 6.30. *Let $n \in \mathbb{N}$. Then the assignment $\mathbf{U} \mapsto [\mathbf{U}]$ induces a bijection between the subsets $\mathbf{U}$ of* $\operatorname{Ind} R$ *with $\sum_{M\in\mathbf{U}} \ell_{\mathrm{end}}(M) = n$ and the fp-idempotent ideals in* $\operatorname{mod} R$ *of length n.*

PROOF. We use the bijection $\mathbf{U} \to [\operatorname{Prod}\mathbf{U}]$ between Ziegler-closed subsets of $\operatorname{Ind} R$ and fp-idempotent ideals in $\operatorname{mod} R$ from Corollary 5.14. It follows from

Lemma 6.27 in combination with the duality between $C(R^{\mathrm{op}})$ and $C(R)$ that the length of $[\operatorname{Prod}\mathbf{U}]$ is precisely the length of $\operatorname{Hom}_{R^{\mathrm{op}}}(R,-)$ in $C(R)/\mathcal{S}$ where $\mathcal{S}$ denotes the Serre subcategory of $C(R)$ corresponding to $\mathbf{U}$. Lemma 6.12 implies that this number equals the endolength of $\prod_{M\in\mathbf{U}} M$ which is $\sum_{M\in\mathbf{U}} \ell_{\mathrm{end}}(M)$. □

COROLLARY 6.31. *Let* $\mathfrak{I}$ *be an fp-idempotent ideal in* $\operatorname{mod} R$ *of length* n. *If* $\mathfrak{I}$ *is nilpotent, then* $\mathfrak{I}^{n+1} = 0$.

CHAPTER 7

Krull-Gabriel dimension

In his thèse [**26**], Gabriel introduced a dimension for every Grothendieck category which he called Krull dimension. In this chapter we shall work with a finitely presented version of Gabriel's Krull dimension. There is also a local variant of that dimension which is defined for any modular lattice and therefore also for the lattice of subobjects of any object in an abelian category. We refer to the appendix for the precise definitions of these dimensions; here they are used to study the complexity of a module category or the complexity of a single module.

7.1. The Krull-Gabriel dimension of a ring

In [**38**], Gruson and Jensen assign to every skeletally small abelian category $\mathcal{C}$ a dimension $\dim \mathcal{C}$. Our aim is to study some of the properties of a module category $\operatorname{Mod} R$ which are related to $\dim C(R)$. Following Geigle, we use the term "Krull-Gabriel dimension" [**32**]. By abuse of terminology, we call $\dim C(R)$ the *Krull-Gabriel dimension* of the ring R and denote it by $\operatorname{KGdim} R$. In fact, $\operatorname{KGdim} R$ is the finitely presented version of Gabriel's Krull dimension of the Grothendieck category $D(R) = (\mathcal{C}^{\mathrm{op}}, \mathrm{Ab})$ where $\mathcal{C} = (\operatorname{mod} R^{\mathrm{op}})^{\mathrm{op}}$. Thus one should speak about the Krull-Gabriel dimension of $(\operatorname{mod} R^{\mathrm{op}})^{\mathrm{op}}$ since the Krull dimension of the ring R is the Krull dimension of the Grothendieck category $\operatorname{Mod} R = (R^{\mathrm{op}}, \mathrm{Ab})$.

First, we are interested in a method to compute the Krull-Gabriel dimension of R. To this end we give a description of the lattice $L(\operatorname{Hom}_R(A, -))$ of subobjects in $C(R^{\mathrm{op}})$ for every finitely presented R-module A.

Fix A in $\operatorname{mod} R$. We consider pairs $x = (X, \phi)$ consisting of a module X and a map $\phi\colon A \to X$ in $\operatorname{mod} R$. Given a second pair $y = (Y, \psi)$, we write $y \leq x$ if there is a map $\alpha\colon X \to Y$ with $\psi = \alpha \circ \phi$. We call x and y *equivalent* if $x \leq y$ and $y \leq x$; we shall not distinguish between a pair x and its equivalence class. We denote by L_A the set of equivalence classes of pairs $x = (X, \phi)$; they form a modular lattice with a unique minimal element $0 = (0, 0)$ and a unique maximal element $1 = (A, \mathrm{id})$. In fact, $x \vee y = (X \coprod Y, \left[\begin{smallmatrix}\phi \\ \psi\end{smallmatrix}\right])$ and $x \wedge y = (Z, \rho)$ where $[\sigma\ \tau]\colon X \coprod Y \to Z$ denotes the cokernel of $\left[\begin{smallmatrix}\phi \\ \psi\end{smallmatrix}\right]$ and $\rho = \sigma \circ \phi$.

LEMMA 7.1. *$L_A \to L(\operatorname{Hom}_R(A, -))$, $(X, \phi) \mapsto \operatorname{Im} \operatorname{Hom}_R(\phi, -)$, is an isomorphism.*

PROOF. Clear. □

Now we can give an alternative description of the Krull-Gabriel dimension of R.

PROPOSITION 7.2. *The following are equivalent for every ordinal α:*

(1) $\operatorname{KGdim} R \leq \alpha$.

(2) $\dim L_X \leq \alpha$ *for every finitely presented* R*-module* X.

(3) $\dim L_R \leq \alpha$.

If R *is right artinian, then* (1) - (3) *are equivalent to:*

(4) $\dim L_S \leq \alpha$ *for every simple* R*-module* S.

PROOF. Let $F \in C(R^{\mathrm{op}})$. Choosing exact sequences $\mathrm{Hom}_R(X, -) \to F \to 0$ and $R^n \to X \to 0$, it is immediately clear that F is a subquotient of $\mathrm{Hom}_R(R, -)^n$. The assertion of the proposition is therefore a consequence of Lemma B.9 and Lemma B.10. □

We include a criterion for $\mathrm{KGdim}\, R = \infty$ which is due to Prest [**64**].

PROPOSITION 7.3. *Let* R *be right artinian. Suppose there exists a family of maps* $(\phi_{ij})_{i \leq j}$ $(i, j \in [0, 1] \cap \mathbb{Q})$ *in the Jacobson radical of* $\mathrm{mod}\, R$ *such that* $\phi_{jk} \circ \phi_{ij} = \phi_{ik}$ *for all* $i \leq j \leq k$ *and* $\phi_{01} \neq 0$. *Then* $\mathrm{KGdim}\, R = \infty$.

PROOF. Let $\phi_{ij} \colon X_i \to X_j$ $(i, j \in [0, 1] \cap \mathbb{Q})$ be the family of maps in $\mathrm{mod}\, R$ and choose $\psi \colon R \to X_0$ with $\phi_{01} \circ \psi \neq 0$. Then the pairs $(X_i, \phi_{0i} \circ \psi)$, $i \in [0, 1] \cap \mathbb{Q}$, form a chain in L_R which is dense since the ϕ_{ij} belong to the Jacobson radical of $\mathrm{mod}\, R$. It follows from Proposition 7.2 and Lemma B.8 that $\mathrm{KGdim}\, R = \dim L_R = \infty$. □

We give some examples.

EXAMPLE 7.4. (1) $\mathrm{KGdim}\, R = 0$ if and only if R is of finite representation type, i.e. R is right artinian and there are only finitely many isomorphism classes of finitely presented indecomposable R-modules [**3**].

(2) A ring R is right pure semi-simple, i.e. every pure-exact sequence in $\mathrm{Mod}\, R$ splits, if and only if $D(R)$ is locally noetherian, c.f. Proposition A.11. In this case $\mathrm{KGdim}\, R < \infty$.

(3) $\mathrm{KGdim}\, R = 1$ is impossible for any artin algebra [**51, 41**].

(4) $\mathrm{KGdim}\, R = 2$ holds for any tame hereditary artin algebra [**32**] or for any Dedekind domain [**43**].

(5) If R is a finite dimensional algebra over a field, then several examples suggest that $\mathrm{KGdim}\, R < \infty$ is closely related to R being of tame domestic representation type. An example of a tame algebra with $\mathrm{KGdim}\, R = \infty$ is given in the proof of Proposition 8.15.

(6) $\mathrm{KGdim}\, R = \infty$ for any finite dimensional algebra R which is of wild representation type; see Proposition 8.15.

(7) Let R be right noetherian. Then the Krull dimension of R in the sense of Gabriel [**26**], i.e. the Krull dimension of the category $\mathrm{Mod}\, R$, equals $\dim \mathrm{mod}\, R$, see Lemma B.5.

Given an epimorphism $R \to S$ we call S a *quotient* of R. It is well-known that the quotient of a representation finite algebra is again representation finite, see for instance [**28**]. This is a special case of the following result since R is of finite representation type if and only if $\mathrm{KGdim}\, R = 0$.

PROPOSITION 7.5. *Let* $R \to S$ *be an epimorphism. Then the Krull-Gabriel dimension of* S *is bounded by the Krull-Gabriel dimension of* R.

PROOF. A homomorphism $f \colon R \to S$ induces an exact functor $f' \colon C(R) \to C(S)$, and it can be shown that f' induces an equivalence $C(R)/\mathcal{S} \to C(S)$ for $\mathcal{S} = \mathrm{Ker}\, f'$ if f is an epimorphism, see Proposition 11.17. The assertion now follows from Lemma B.1. □

7.2. The Krull-Gabriel dimension of a module

Viewing the Krull-Gabriel dimension of a ring R as a global invariant of $\operatorname{Mod} R$ it is also posssible to define in a similar way for every R-module a local invariant. Let M be an R-module and denote by $\operatorname{Latt}(M)$ the lattice of subgroups of finite definition of M. We call $\dim \operatorname{Latt}(M)$ the *Krull-Gabriel dimension* of the module M and denote it by $\operatorname{KGdim} M$. Note that the Krull-Gabriel dimension of the ring R is usually different from the Krull-Gabriel dimension of the module R. We collect some basic properties of this dimension.

PROPOSITION 7.6. *Let M be an R-module.*

(1) $\operatorname{KGdim} M = \dim \mathcal{E}_M$ *where $\mathcal{E}_M$ denotes the endocategory of M.*
(2) $\operatorname{KGdim} M$ *is bounded by the Krull-Gabriel dimension of the ring R.*
(3) *If $0 \to M' \to M \to M'' \to 0$ is a pure-exact sequence, then* $\operatorname{KGdim} M = \sup(\operatorname{KGdim} M', \operatorname{KGdim} M'')$.

PROOF. (1) Combine Lemma 6.1 and Lemma B.9.

(2) The endocategory of M is equivalent to a quotient category of $C(R)^{\mathrm{op}}$. Using (1) and Lemma B.1 we obtain $\operatorname{KGdim} R \geq \operatorname{KGdim} M$.

(3) This is a consequence of the following lemma. □

Given a map $\phi\colon R \to X$ in $\operatorname{mod} R^{\mathrm{op}}$, we denote by M_ϕ the kernel of the map $M \to M \otimes_R X$, $m \mapsto m \otimes \phi(1)$.

LEMMA 7.7. *Let $0 \to M' \to M \to M'' \to 0$ be a pure-exact sequence. Then*

$$\operatorname{Latt}(M) \longrightarrow \operatorname{Latt}(M') \times \operatorname{Latt}(M''), \quad M_\phi \mapsto (M'_\phi, M''_\phi)$$

is an injective lattice homomorphism.

PROOF. For every map ϕ the induced sequence $0 \to M'_\phi \to M_\phi \to M''_\phi \to 0$ is exact. Therefore the map is well-defined since $M'_\phi = M' \cap M_\phi$, and the injectivity follows easily. □

EXAMPLE 7.8. (1) If M is a Σ-pure-injective module, then $\operatorname{KGdim} M < \infty$. A non-zero module M is endofinite if and only if $\operatorname{KGdim} M = 0$.

(2) Let $M \neq 0$ be a module over a noetherian algebra R such that $\operatorname{KGdim} M < \infty$. If M has no finitely presented indecomposable direct summand, then there exists a generic R-module [**51**, Theorem 5.6].

(3) Let R be right coherent, i.e. the category $\operatorname{mod} R$ is abelian. Then the dimension $\dim \operatorname{mod} R$ equals the Krull-Gabriel dimension of the R^{op}-module R. Therefore the Krull dimension of the ring R in the sense of Gabriel [**26**] equals the Krull-Gabriel dimension of the R^{op}-module R provided that R is right noetherian.

We devote the rest of this section to a discussion of pure-injective modules having Krull-Gabriel dimension. The basic structure theorem goes as follows.

THEOREM 7.9. *Let M be a pure-injective R-module. Suppose that* $\operatorname{KGdim} M < \infty$. *Then there is a family of modules $(N_i)_{i\in I}$ in $\operatorname{Ind} R$ such that M is the pure-injective envelope of $\coprod_{i\in I} N_i$. Moreover, given a family $(L_j)_{j\in J}$ in $\operatorname{Ind} R$, then $\coprod_{i\in I} N_i$ and $\coprod_{j\in J} L_j$ have the same pure-injective envelope if and only if there exists a bijection $\pi\colon I \to J$ such that $L_{\pi(i)} = N_i$ for all $i \in I$.*

PROOF. The assertion of this theorem is a direct consequence of a result from [**26**] about injective objects in Grothendieck categories having Krull dimension. Let $\mathcal{S}_M = \{F \in C(R) \mid \operatorname{Hom}(F, T_M) = 0\}$ and let $\mathcal{T}_M = \varinjlim \mathcal{S}_M$. Then the Krull-dimension of $D(R)/\mathcal{T}_M$ is bounded by $\operatorname{KGdim} M$. This follows from Lemma B.5 and the fact that $\operatorname{KGdim} M = \dim C(R)/\mathcal{S}_M$. Therefore T_M is an injective object in a Grothendieck category having Krull dimension and the assertion can be derived from [**26**, II, Théorème 1]. □

We are now in a position to apply the duality map $\operatorname{Ref} R \to \operatorname{Ref} R^{\mathrm{op}}$, $M \mapsto M^{\vee}$, between the pure-reflexive modules over R and R^{op}, respectively.

COROLLARY 7.10. *The assignment $M \mapsto M^{\vee}$ induces for every ordinal α a bijection between the isomorphism classes of pure-injective R-modules M such that* $\operatorname{KGdim} M = \alpha$ *and the isomorphism classes of pure-injective R^{op}-modules N such that* $\operatorname{KGdim} N = \alpha$.

PROOF. Let M be a pure-injective module with $\operatorname{KGdim} M < \infty$. We need to show that M is pure-reflexive. This follows from Lemma 4.10 if M is indecomposable, and the general case then follows with Theorem 7.9. Observe that M and $M^{\vee}$ are purely opposed by Proposition 4.16. This implies $\operatorname{KGdim} M = \operatorname{KGdim} M^{\vee}$, and the assertion is therefore a consequence of Theorem 4.14. □

CHAPTER 8

The infinite radical

In this chapter we introduce a new radical series of the category $\operatorname{mod} R$ of finitely presented R-modules. This radical series can be used to extend the preinjective partition of $\operatorname{mod} R$ which Auslander and Smalø introduced in [**7**]. We discuss also the relation between this new radical series and the powers of the Jacobson radical of $\operatorname{mod} R$ which have been studied by Prest [**64**].

8.1. The preinjective dimension of a module

In their study of the representation theory of finite-dimensional tensor algebras Dlab and Ringel described certain modules which they called preprojective and preinjective modules [**23**]. Later Auslander and Smalø defined these concepts for arbitrary artin algebras [**7**]. Our aim is to extend the class of preinjective modules which Auslander and Smalø introduced. To this end we define a radical series $(\mathfrak{rad}_\alpha)_\alpha$ of the category $\operatorname{mod} R$. Using this radical series, we assign to every finitely presented module a preinjective dimension. For an artin algebra, it will be shown that this dimension is finite if and only if the module is preinjective in the sense of Auslander and Smalø. Also, we shall see that the preinjective dimension leads to a refinement of the Krull-Gabriel dimension. We mention that our approach is different from that of Zimmermann-Huisgen in her work on so-called strong preinjective partitions [**82**].

Fix a ring R and recall that for every functor F in $C(R^{\mathrm{op}})$ there is defined a dimension $\dim F$. We obtain a descending chain of ideals in $\operatorname{mod} R$ if we define for every ordinal α

$$\mathfrak{rad}_\alpha = \bigcap_{\dim F \leq \alpha} \operatorname{ann} F$$

where F runs through all F in $C(R^{\mathrm{op}})$. Using this *radical series* $(\mathfrak{rad}_\alpha)_\alpha$ we define for every finitely presented R-module M a *socle series* $(\operatorname{soc}_\alpha M)_\alpha$ as follows:

$$\operatorname{soc}_\alpha M = \bigcap_{\phi \in \mathfrak{rad}_\alpha} \operatorname{Ker} \phi$$

where ϕ runs through all maps starting at M. The *preinjective dimension* $\operatorname{pidim} M$ of a finitely presented non-zero module M is the least ordinal (or ∞ if such an ordinal does not exist) such that $\operatorname{soc}_\alpha M \neq 0$. For $M = 0$ let $\operatorname{pidim} M = 0$. We collect some basic facts which are direct consequences of the definitions.

LEMMA 8.1. *The following holds:*

(1) $\mathfrak{rad}_\alpha = \bigcap_{\beta<\alpha} \mathfrak{rad}_\beta$ *for every limit ordinal* α.
(2) $\operatorname{soc}_\alpha M = \bigcup_{\beta<\alpha} \operatorname{soc}_\beta M$ *for every limit ordinal* α.
(3) *If* $\operatorname{pidim} M < \infty$, *then* $\operatorname{pidim} M$ *is not a limit ordinal.*

PROOF. (1) follows from the fact that $\dim F$ is not a limit ordinal for every $F \in C(R^{\mathrm{op}})$ provided that $\dim F < \infty$. (2) follows from (1), and (3) follows from (2). □

PROPOSITION 8.2. *Let R be a ring of Krull-Gabriel dimension α. Then there exists $n \in \mathbb{N}$ such that $\mathfrak{rad}_{\omega\alpha+n} = 0$ and therefore $\operatorname{pidim} M \leq \omega\alpha + n$ for every M in $\operatorname{mod} R$.*

PROOF. Let $\mathcal{C} = C(R^{\mathrm{op}})$. If $\operatorname{KGdim} R = \alpha$, then $\dim \operatorname{Hom}_{R^{\mathrm{op}}}(R, -) = n$ in $\mathcal{C}/\mathcal{C}_{\alpha-1}$ for some $n \in \mathbb{N}$, and therefore $\dim F \leq \omega\alpha+n$ for all $F \in \mathcal{C}$ by Lemma B.10. Thus $\mathfrak{rad}_{\omega\alpha+n} = 0$ and $\operatorname{pidim} M \leq \omega\alpha + n$ for all M follows. □

From now on we assume for the rest of this section that R is an artin algebra. We collect some basic properties of the radical series $(\mathfrak{rad}_\alpha)_\alpha$. To this end recall that the *Jacobson radical* $\operatorname{rad} \mathcal{C}$ of a skeletally small preadditive category $\mathcal{C}$ is the intersection of all maximal ideals in $\mathcal{C}$. Given $F \in (\mathcal{C}, \mathrm{Ab})$, we denote by $\operatorname{rad} F$ the intersection of all maximal subobjects of F, and $\operatorname{rad}^n F = \operatorname{rad}(\operatorname{rad}^{n-1} F)$ for every $n \in \mathbb{N}$ where $\operatorname{rad}^0 F = F$.

LEMMA 8.3. *Let $F \in C(R^{\mathrm{op}})$ and $n \in \mathbb{N}_0$. Then the following are equivalent:*

(1) $\dim F \leq n$.
(2) $\operatorname{rad}^n F = 0$.
(3) *There is an epimorphism* $\operatorname{Hom}_R(X, -)/\operatorname{rad}^n \operatorname{Hom}_R(X, -) \to F$ *for some* X *in* $\operatorname{mod} R$.

PROOF. Use the fact that every simple functor in $(\operatorname{mod} R, \mathrm{Ab})$ is finitely presented [**6**]. □

PROPOSITION 8.4. *The following holds:*

(1) $\mathfrak{rad}_1$ *is the Jacobson radical of* $\operatorname{mod} R$.
(2) $\mathfrak{rad}_n(X, Y) = \operatorname{rad}^n \operatorname{Hom}(X, -)(Y)$ *for all* $X, Y \in \operatorname{mod} R$ *and* $n \in \mathbb{N}_0$.

PROOF. (1) Any ideal in $\operatorname{mod} R$ is maximal if and only if it is of the form $\operatorname{ann} S$ for some simple functor $S \in (\operatorname{mod} R, \mathrm{Ab})$. The simple functors are finitely presented [**6**] and belong therefore to $C(R^{\mathrm{op}})$.

(2) Use the preceding lemma. □

We shall need the following lemma.

LEMMA 8.5. *The following are equivalent for M in $\operatorname{mod} R$ and an ordinal α:*

(1) $\operatorname{pidim} M < \omega(\alpha + 1)$.
(2) *There exists a subfunctor $F \subseteq \operatorname{Hom}_R(S, -)$ in $C(R^{\mathrm{op}})$ for some simple R-module S such that $F(M) \neq 0$ and $\dim F < \omega(\alpha + 1)$.*

PROOF. (1) $\Rightarrow$ (2) Choose a simple submodule $S \subseteq \operatorname{soc}_{\omega(\alpha+1)} M$ and denote by F the image of the induced map $\operatorname{Hom}_R(M, -) \to \operatorname{Hom}_R(S, -)$. Using Yoneda's lemma it follows that $F(\phi) = 0$ for all $\phi \in \mathfrak{rad}_{\omega(\alpha+1)}$, and therefore $F \in C(R^{\mathrm{op}})_\alpha$ by Corollary 5.9 since $\mathfrak{rad}_{\omega(\alpha+1)} = \operatorname{ann} C(R^{\mathrm{op}})_\alpha$. Clearly, $F(M) \neq 0$ and $\dim F < \omega(\alpha + 1)$ follows from Lemma B.3.

(2) $\Rightarrow$ (1) A subobject $F \subseteq \operatorname{Hom}_R(S, -)$ with $F(M) \neq 0$ induces a non-zero map $S \to M$. If $\dim F \leq \omega\alpha + n$ for some $n \in \mathbb{N}$, then $S \subseteq \operatorname{soc}_{\omega\alpha+n} M$ and therefore $\operatorname{pidim} M < \omega(\alpha + 1)$. □

We formulate now the main result of this section. It shows that the preinjective dimension is a refinement of the Krull-Gabriel dimension.

THEOREM 8.6. *Let R be an artin algebra. Then the following are equivalent for every ordinal α:*

(1) $\operatorname{KGdim} R \leq \alpha$.
(2) *There exist $n \in \mathbb{N}$ such that* $\mathfrak{rad}_{\omega\alpha+n} = 0$.
(3) *There exist $n \in \mathbb{N}$ such that* $\operatorname{pidim} M \leq \omega\alpha + n$ *for every M in* $\operatorname{mod} R$.
(4) *There exist $n \in \mathbb{N}$ such that* $\operatorname{pidim} S \leq \omega\alpha + n$ *for every simple S in* $\operatorname{mod} R$.

PROOF. (1) $\Rightarrow$ (2) $\Rightarrow$ (3) $\Rightarrow$ (4) are covered by Proposition 8.2. Therefore assume (4). It follows from Lemma 8.5 that $\operatorname{Hom}_R(S, -) \in C(R^{\mathrm{op}})_\alpha$ for every simple R-module S. Thus $C(R^{\mathrm{op}})_\alpha = C(R^{\mathrm{op}})$ since $C(R^{\mathrm{op}})_\alpha$ is a Serre subcategory and $C(R^{\mathrm{op}})$ is generated by the representable functors corresponding to the simple R-modules. We conclude that $\operatorname{KGdim} R \leq \alpha$. □

We obtain the following relation between the preinjective dimension and the concept introduced by Auslander and Smalø.

PROPOSITION 8.7. *A finitely presented indecomposable R-module M is preinjective in the sense of Auslander and Smalø if and only if* $\operatorname{pidim} M$ *is finite.*

PROOF. Combine the characterization in [**7**, Theorem 5.2] with Lemma 8.5. □

EXAMPLE 8.8. (1) If M is indecomposable injective with simple submodule S, then $\operatorname{soc}_1 M = S$.

(2) Let R be a tame hereditary artin algebra. It is well-known that the finitely presented indecomposable R-modules fall into three classes: preinjective, regular, and preprojective R-modules [**22**]. Let M be a finitely presented indecomposable R-module. Then M is preinjective if and only if $\operatorname{pidim} M < \omega$; M is regular if and only if $\omega < \operatorname{pidim} M < \omega 2$; M is preprojective if and only if $\omega 2 < \operatorname{pidim} M < \omega 3$.

8.2. Transfinite powers of the Jacobson radical

Throughout this section R denotes an artin algebra. We denote by $\mathfrak{rad} = \operatorname{rad}(\operatorname{mod} R)$ the *Jacobson radical* of $\operatorname{mod} R$, i.e. the intersection of all maximal ideals in $\operatorname{mod} R$. There are various ways to define the power $\mathfrak{rad}^\alpha$ for every ordinal α, and we shall follow Prest [**64**] who gave a definition which is similar to that of Krause and Lenagan [**46**]. After discussing some basic properties of this radical series $(\mathfrak{rad}^\alpha)_\alpha$ it will be shown that $\mathfrak{rad}^\alpha$ is closely related to $\mathfrak{rad}_\alpha$. In particular, $\mathfrak{rad}^\alpha = 0$ for some ordinal α if the Krull-Gabriel dimension of R is an ordinal.

We begin with the definition of $\mathfrak{I}^\alpha$ for any ideal $\mathfrak{I}$ and any ordinal α. If α is a finite ordinal, then $\mathfrak{I}^\alpha = \{\sum_{i=1}^r x_{i1} \cdots x_{i\alpha} \mid x_{ij} \in \mathfrak{I}, r \in \mathbb{N}\}$. If α is a limit ordinal, let $\mathfrak{I}^\alpha = \bigcap_{\gamma<\alpha} \mathfrak{I}^\gamma$, and if α is an infinite non-limit ordinal, so uniquely of the form $\alpha = \beta + n$ for some limit ordinal β and some natural number $n \geq 1$, let $\mathfrak{I}^\alpha = (\mathfrak{I}^\beta)^{n+1}$. Finally, we define $\mathfrak{I}^\infty = \bigcap_\alpha \mathfrak{I}^\alpha$ where α runs through all ordinals.

The lattice of ideals in $\operatorname{mod} R$ has the following interesting finiteness property. This generalizes an important observation of Schröer [**77**].

PROPOSITION 8.9. *Suppose there are up to isomorphism n simple R-modules. If $(\mathfrak{I}_i)_{i\in I}$ is a family of ideals in* $\operatorname{mod} R$ *satisfying $\bigcap_{i\in I} \mathfrak{I}_i = 0$, then there are $i_1, \ldots, i_n$ in I such that $\bigcap_{j=1}^n \mathfrak{I}_{i_j} = 0$.*

PROOF. Denote for every simple R-module S by ϕ_S the composition of the projective cover $P \to S$ with the injective envelope $S \to I$. The assertion follows immediately from the fact that $\mathfrak{I} \neq 0$ for any ideal $\mathfrak{I}$ if and only if $\phi_S \in \mathfrak{I}$ for some simple S. □

The following proposition collects some of the well-known properties of the radical series $(\mathfrak{rad}^\alpha)_\alpha$. The second property shows that the usual definition of $\mathfrak{rad}^\alpha$ is not of interest if $\alpha > \omega$.

PROPOSITION 8.10. *The following holds:*

(1) $\mathfrak{rad}^n(X,Y) = \operatorname{rad}^n \operatorname{Hom}(X,-)(Y)$ *for all* $X, Y \in \operatorname{mod} R$ *and* $n \in \mathbb{N}_0$.
(2) $\mathfrak{rad}(\mathfrak{rad}^\omega) = \mathfrak{rad}^\omega = (\mathfrak{rad}^\omega)\mathfrak{rad}$.
(3) $\mathfrak{rad}^\infty \neq 0$ *if and only if there exists a non-zero idempotent ideal in* $\operatorname{mod} R$ *which is contained in* $\mathfrak{rad}$.
(4) *If* $\mathfrak{rad}^\infty = 0$, *then there exists an ordinal* α *such that* $\mathfrak{rad}^\alpha \neq 0$ *and* $\mathfrak{rad}^{\alpha+1} = 0$.

PROOF. (1) We use induction on n. The assertion is clear for $n = 0, 1$. For $n > 1$ choose a projective cover $\operatorname{Hom}_R(X',-) \to \operatorname{rad}^{n-1} \operatorname{Hom}_R(X,-)$ which induces a map $\rho\colon X \to X'$. We obtain the following commutative diagram with exact rows and columns:

$$\begin{array}{ccccccc}
0 \longrightarrow & \operatorname{rad}\operatorname{Hom}_R(X',-) & \longrightarrow & \operatorname{Hom}_R(X',-) & \longrightarrow & F & \longrightarrow 0 \\
& \downarrow & & \downarrow & & \| & \\
0 \longrightarrow & \operatorname{rad}^n \operatorname{Hom}_R(X,-) & \longrightarrow & \operatorname{rad}^{n-1}\operatorname{Hom}_R(X,-) & \longrightarrow & F & \longrightarrow 0 \\
& \downarrow & & \downarrow & & & \\
& 0 & & 0 & & &
\end{array}$$

If $\phi \in \mathfrak{rad}^n(X,Y)$, then we may assume that $\phi = \phi'' \circ \phi'$ with $\phi' \in \mathfrak{rad}^{n-1}$ and $\phi'' \in \mathfrak{rad}$. By assumption, $\phi' = \psi \circ \rho$ for some map ψ, and $\phi'' \circ \psi \in \operatorname{rad}\operatorname{Hom}_R(X',-)(Y)$. The commutativity of the diagram implies $\phi \in \operatorname{rad}^n \operatorname{Hom}_R(X,-)(Y)$. The proof for $\operatorname{rad}^n \operatorname{Hom}(X,-)(Y) \subseteq \mathfrak{rad}^n(X,Y)$ is analogous.

(2) We use the fact that for every pair $X, Y \in \operatorname{mod} R$ there exists $n \in \mathbb{N}$ such that $\mathfrak{rad}^\omega(X,Y) = \mathfrak{rad}^m(X,Y)$ for every $m \geq n$ in $\mathbb{N}$. Now assume, without loss of generality, that Y is indecomposable. Choose a right almost split map $Y' \to Y$ and suppose that $\mathfrak{rad}^\omega(X,Y') = \mathfrak{rad}^{n'}(X,Y')$. For $m = n + n'$ we obtain

$$\mathfrak{rad}^\omega(X,Y) = \mathfrak{rad}^m(X,Y) = \mathfrak{rad}(Y',Y)\mathfrak{rad}^{m-1}(X,Y') = \mathfrak{rad}(Y',Y)\mathfrak{rad}^\omega(X,Y').$$

Thus $\mathfrak{rad}^\omega = \mathfrak{rad}(\mathfrak{rad}^\omega)$, and the proof of $\mathfrak{rad}^\omega = (\mathfrak{rad}^\omega)\mathfrak{rad}$ is analogous.

(3) By definition, $\mathfrak{rad}^\infty$ is idempotent. If $\mathfrak{I}^2 = \mathfrak{I} \subseteq \mathfrak{rad}$, then $\mathfrak{I} \subseteq \mathfrak{rad}^\infty$.

(4) Use Proposition 8.9. □

Our next aim is to compare $\mathfrak{rad}^\alpha$ with $\mathfrak{rad}_\alpha$. We need the following lemma.

LEMMA 8.11. *Let $\mathcal{S}$ be a Serre subcategory of $C(R^{\mathrm{op}})$ containing all finite length objects. Let $F \in C(R^{\mathrm{op}})$ and suppose that* $\dim F = n \in \mathbb{N}$ *in* $C(R^{\mathrm{op}})/\mathcal{S}$. *Then* $(\operatorname{ann} \mathcal{S})^{2n} \subseteq \operatorname{ann} F$.

PROOF. We use the notation which was introduced before Lemma 5.4. Suppose first that F is simple. We observe that F/tF is not finitely presented since every finitely presented object has a simple subobject which belongs to $\mathcal{S}$ by assumption. Now we claim that $t'(F/tF)^* = (F/tF)^*$. To this end assume $t'(F/tF)^* \neq (F/tF)^*$.

Thus there exists a finitely presented subobject of $(F/tF)^*$ which does not belong to $\mathcal{S}^*$. Applying the duality we obtain a proper quotient object F'' of F/tF which does not belong to $\mathcal{S}$. If $F'' = F/F'$, then F' properly contains tF but belongs to $\mathcal{S}$ since F and therefore F'' are simple in $C(R^{\mathrm{op}})/\mathcal{S}$. This contradiction proves $t'(F/tF)^* = (F/tF)^*$. Therefore $\operatorname{ann}\mathcal{S}$ annihilates tF and F/tF, and $(\operatorname{ann}\mathcal{S})^2 \subseteq \operatorname{ann} F$ follows with Lemma C.8. Using induction on $n = \dim F$, one obtains $(\operatorname{ann}\mathcal{S})^{2n} \subseteq \operatorname{ann} F$. □

We are now in a position to prove the following relation between $\mathfrak{rad}^\alpha$ and $\mathfrak{rad}_\alpha$.

THEOREM 8.12. *Let α be any ordinal, written uniquely as $\alpha = \omega\beta + n$ with $n < \omega$. Then $\mathfrak{rad}^{\alpha+n} \subseteq \mathfrak{rad}_\alpha$. More precisely, the following holds:*

(1) *If $\alpha \leq \omega$, then $\mathfrak{rad}^\alpha = \mathfrak{rad}_\alpha$.*
(2) *If $n = 0$, then $\mathfrak{rad}^\alpha \subseteq \mathfrak{rad}_\alpha$.*
(3) *If $n > 0$, then $\mathfrak{rad}^{\alpha+(n-1)} \subseteq \mathfrak{rad}_\alpha$.*

PROOF. (1) follows from Propositions 8.4 and 8.10. (2) and (3) follow by transfinite induction. For a non-limit ordinal α one observes that $\mathfrak{rad}^{\alpha+(n-1)} = (\mathfrak{rad}^{\omega\beta})^{2n}$ and uses Lemma 8.11. For a limit ordinal α one uses that $\mathfrak{rad}^\alpha = \bigcap_{\gamma<\alpha} \mathfrak{rad}^\gamma$ and $\mathfrak{rad}_\alpha = \bigcap_{\gamma<\alpha} \mathfrak{rad}_\gamma$. □

We discuss some consequences. Recall that for any class $\mathcal{C}$ of R-modules we denote by $[\mathcal{C}]$ the ideal of maps in $\operatorname{mod} R$ which factor through a finite coprododuct of modules in $\mathcal{C}$.

COROLLARY 8.13. *The following are equivalent for a class $\mathcal{C}$ of R-modules:*

(1) *There is no module in $\mathcal{C}$ having a finitely presented indecomposable direct summand.*
(2) *$[\mathcal{C}] \subseteq \mathfrak{rad}$.*
(3) *$[\mathcal{C}] \subseteq \mathfrak{rad}^\omega$.*

Moreover, if $\mathcal{X}$ denotes the definable subcategory of all R-modules having no finitely presented indecomposable direct summand, then $[\mathcal{X}] = \mathfrak{rad}^\omega$.

PROOF. The ideal $\mathfrak{rad}_\omega$ is precisely the fp-idempotent ideal which corresponds to the definable subcategory $\mathcal{X}$ of all R-modules which have no finitely presented indecomposable direct summand. Thus $[\mathcal{X}] = \mathfrak{rad}^\omega$ by Theorem 8.12. The equivalence of the conditions (1) – (3) now follows since (1) holds if and only if $\mathcal{C} \subseteq \mathcal{X}$. □

COROLLARY 8.14. *If R has Krull-Gabriel dimension α, then $\mathfrak{rad}^{\omega\alpha+n} = 0$ for some $n \in \mathbb{N}$.*

PROOF. Combine Proposition 8.2 and Theorem 8.12. □

It is an open question whether $\mathfrak{rad}^\alpha = 0$ for some ordinal α implies that $\operatorname{KGdim} R < \infty$. Notice that $\mathfrak{rad}^\infty \neq 0$ and $\operatorname{KGdim} R = \infty$ for any algebra R which is of wild representation type. In particular, any algebra R such that $\mathfrak{rad}^\omega$ is nilpotent needs to be tame [**44**].

PROPOSITION 8.15. *Let R be a finite dimensional algebra over some field k and suppose that R has wild representation type. Then $\mathfrak{rad}^\infty \neq 0$ and $\operatorname{KGdim} R = \infty$.*

PROOF. Let $S = k[X,Y]/(XY, X^2, Y^3)$. Note that this algebra is of tame representation type. In fact, the classification of the finite dimensional indecomposable S-modules is well-known, and there exists also a convenient description of the morphism space between two indecomposables [**15**]. Using this knowledge, it is not hard to find a family M_i, $i \in I = \{m \cdot 2^{-n} \in [0,1] \mid m \in \mathbb{Z}, n \in \mathbb{N}\}$, of indecomposables in $\operatorname{mod} S$ and a family of non-zero maps $\phi_{ij}\colon M_i \to M_j$ for each pair $i \leq j$ in I such that $\phi_{ik} = \phi_{jk} \circ \phi_{ij}$ and $\phi_{ij} \in \mathfrak{rad}_S = \operatorname{rad}(\operatorname{mod} S)$ for all $i \leq j \leq k$. Now let $\mathfrak{I}$ be the ideal in $\operatorname{mod} S$ which is generated by the ϕ_{ij}. Clearly, $\mathfrak{I}^2 = \mathfrak{I}$ and therefore $\mathfrak{rad}_S^\infty \neq 0$ by Proposition 8.10. Now let $\mathfrak{rad}_R = \operatorname{rad}(\operatorname{mod} R)$. By definition, there exists a representation embedding $f\colon \operatorname{mod} S \to \operatorname{mod} R$ since R is of wild representation type [**14**]. The definition of a representation embedding implies that f is faithful with $f(\mathfrak{rad}_S) \subseteq \mathfrak{rad}_R$. Thus $\mathfrak{rad}_R^\infty \neq 0$, and $\operatorname{KGdim} R = \infty$ then follows from Corollary 8.14. □

We end this section with an example which is due to Schröer [**77**]. Recall that an artin algebra R is *domestic* if there are, up to isomorphism, only finitely many generic R-modules.

EXAMPLE 8.16. Let α be an ordinal of the form $\alpha = \omega\beta + n$ with $(1,1) \neq (\beta, n) \in \mathbb{N} \times \mathbb{N}$. Then there exists a domestic string algebra R_α such that $\mathfrak{rad}^\alpha = 0$ and $\mathfrak{rad}^{\alpha-1} \neq 0$ where $\mathfrak{rad} = \operatorname{rad}(\operatorname{mod} R_\alpha)$. Therefore there exists for every $n \in \mathbb{N}$ a domestic algebra R with $\operatorname{KGdim} R \geq n$.

CHAPTER 9

Functors between module categories

Functors between module categories help to understand the properties of a fixed module category. In this chapter we concentrate on a class of functors which arises frequently and preserves a number of finiteness conditions.

9.1. Coherent functors

We characterize the functors which commute with direct limits and products, and we present some of their basic properties. Most of this material is taken from [**54, 52**].

THEOREM 9.1. *The following are equivalent for a functor* $f\colon \operatorname{Mod} S \to \operatorname{Mod} R$*:*

(1) *The functor* f *commutes with direct limits and products.*
(2) *There is an exact functor* $f'\colon C(R) \to C(S)$ *and a functorial isomorphism* $E_{f(M)} \simeq E_M \circ f'$ *for all* $M \in \operatorname{Mod} S$.
(3) *The composition* $F = \operatorname{Hom}_R(R, -) \circ f\colon \operatorname{Mod} S \to \mathrm{Ab}$ *has a presentation* $\operatorname{Hom}_S(Y, -) \to \operatorname{Hom}_S(X, -) \to F \to 0$ *with* $X, Y \in \operatorname{mod} S$*, and there is a ring homomorphism* $\alpha\colon R \to \operatorname{End}(F)^{\mathrm{op}}$ *such that* $m \cdot r = m \cdot \alpha(r)$ *for all* $m \in f(M)$ *and* $r \in R$.

A functor $\operatorname{Mod} S \to \operatorname{Mod} R$ between two module categories is said to be *coherent* if it satisfies one of the equivalent conditions in the preceding theorem. We denote by $C(S, R)$ the category of coherent functors $\operatorname{Mod} S \to \operatorname{Mod} R$ and mention an immediate consequence of the preceding theorem.

COROLLARY 9.2. $C(S, R)$ *is an abelian category which is equivalent to the functor category* $(R^{\mathrm{op}}, C(S^{\mathrm{op}}))$. *Moreover,* $C(S, R)^{\mathrm{op}} \simeq C(S^{\mathrm{op}}, R^{\mathrm{op}})$.

The following lemma is required for the proof of the theorem; it is due to Jensen and Lenzing [**43**].

LEMMA 9.3. *A functor* $\operatorname{Mod} S \to \operatorname{Mod} R$ *commuting with coproducts and products sends pure-injectives to pure-injectives.*

PROOF. A module M is pure-injective if and only if for every set I the summation map $M^{(I)} \to M$ factors through the canonical map $M^{(I)} \to M^I$, see [**43**, Proposition 7.1] or [**52**, Theorem 2.6]. The assertion immediately follows from this characterization. □

PROOF OF THEOREM 9.1. (1) ⇒ (2) We extend f to a functor $f_*\colon D(S) \to D(R)$ as follows. Let X be an object in $D(S)$. Write X as a direct limit of finitely presented functors, i.e. $X = \varinjlim \operatorname{Ker} T_{\phi_i}$ with $\phi_i \in \operatorname{mod} S$ for all i, and define $f_*(X) = \varinjlim \operatorname{Ker} T_{f(\phi_i)}$. It is not hard to check that f_* is left exact and commutes with direct limits and products; in particular $f_*(T_M) = T_{f(M)}$ for all

$M \in \operatorname{Mod} S$. The functor f_* has a left adjoint $f^*: D(R) \to D(S)$ by the Adjoint Functor Theorem [**59**, Corollary V.3.2] which is exact since f_* preserves injectives by the preceding lemma. The functor f^* induces an exact functor $f': C(R) \to C(S)$ by Lemma 1.1 since f_* commutes with direct limits. It is left to the reader to verify that $E_{f(M)} \simeq E_M \circ f'$ for all $M \in \operatorname{Mod} S$.

(2) $\Rightarrow$ (3) Choose a map ϕ in $\operatorname{mod} S$ such that $f'(T_R) = \operatorname{Ker} T_\phi$. It is not hard to check that $F = \operatorname{Hom}_R(R, -) \circ f = \operatorname{Coker} H_\phi$. The duality $C(S) \to C(S^{\mathrm{op}})$ sends $\operatorname{Ker} T_\phi$ to $\operatorname{Coker} H_\phi$, and therefore f' induces a ring homomorphism $\alpha: R \to \operatorname{End}(F)^{\mathrm{op}}$ which gives the R-action on $F(M)$.

(3) $\Rightarrow$ (1) This follows from the well-known fact that a representable functor $\operatorname{Hom}_S(X, -)$ commutes with direct limits and products if X is finitely presented. □

COROLLARY 9.4. *A functor $f: \operatorname{Mod} R \to \operatorname{Mod} S$ is coherent if and only if there exists a functor $f_*: D(S) \to D(R)$ having the following properties:*

(1) *$f_*(T_M) = T_{f(M)}$ for all $M \in \operatorname{Mod} S$.*
(2) *f_* has a left adjoint which is exact and sends finitely presented objects to finitely presented objects.*

EXAMPLE 9.5. (1) A functor $f: \operatorname{Mod} R \to \operatorname{Mod} \mathbb{Z} = \mathrm{Ab}$ is coherent if and only if there exists a presentation $\operatorname{Hom}_R(Y, -) \to \operatorname{Hom}_R(X, -) \to f \to 0$ with $X, Y \in \operatorname{mod} R$. Therefore $C(R^{\mathrm{op}}) \simeq C(R, \mathbb{Z})$.

(2) A tensor functor $- \otimes_S B: \operatorname{Mod} S \to \operatorname{Mod} R$ commutes with direct limits; it commutes with products if and only if B is finitely presented over S. Therefore a functor $f: \operatorname{Mod} S \to \operatorname{Mod} R$ is coherent and right exact if and only if $f \simeq - \otimes_S B$ for some bimodule B which is finitely presented over S.

Now suppose that $f: \operatorname{Mod} S \to \operatorname{Mod} R$ is coherent and choose for every map $\phi \in \operatorname{mod} R$ a map $\phi_f \in \operatorname{mod} S$ such that $f'(\operatorname{Ker} T_\phi) = \operatorname{Ker} T_{\phi_f}$. We write $\Phi_f = \{\phi_f \mid \phi \in \Phi\}$ for every $\Phi \subseteq \operatorname{mod} R$. With this notation we obtain the following consequence of the preceding theorem.

COROLLARY 9.6. *Let $f: \operatorname{Mod} S \to \operatorname{Mod} R$ be a coherent functor and let Φ be a collection of maps in $\operatorname{mod} R$. Then every S-module M is Φ_f-injective if and only if $f(M)$ is Φ-injective, and therefore $f^{-1}((\operatorname{Mod} R)_\Phi) = (\operatorname{Mod} S)_{\Phi_f}$.*

PROOF. We use the pair of adjoint functors between $D(R)$ and $D(S)$ from the preceding proof. We have for each $\phi \in \operatorname{mod} R$ and $M \in \operatorname{Mod} S$ that $f^*(\operatorname{Ker} T_\phi) = \operatorname{Ker} T_{\phi_f}$ and $f_*(T_M) = T_{f(M)}$. The assertion now follows from the adjointness isomorphism $\operatorname{Hom}(\operatorname{Ker} T_{\phi_f}, T_M) \simeq \operatorname{Hom}(\operatorname{Ker} T_\phi, T_{f(M)})$ together with Lemma 2.2. □

We mention some further properties of coherent functors. To this end denote for any module M by $\ell_{\mathrm{end}}(M)$ its endolength.

COROLLARY 9.7. *A coherent functor $f: \operatorname{Mod} S \to \operatorname{Mod} R$ has the following properties:*

(1) *If $\mathbf{U} \subseteq \operatorname{Ind} S$ is Ziegler-closed, then the indecomposable direct summands of modules in $f(\mathbf{U})$ form a Ziegler-closed subset of* $\operatorname{Ind} R$.
(2) *There is $c \in \mathbb{N}$ such that $\ell_{\mathrm{end}}(f(M)) \le c \cdot \ell_{\mathrm{end}}(M)$ for all $M \in \operatorname{Mod} S$.*
(3) *Let $M \in \operatorname{Mod} S$. Then the restriction via the homomorphism $\operatorname{End}_S(M) \to \operatorname{End}_R(f(M))$ induces a faithful and exact functor $\mathcal{E}_{f(M)} \to \mathcal{E}_M$ between the endocategories of M and $f(M)$.*

(4) $\operatorname{KGdim} f(M) \leq \operatorname{KGdim} M$ *for every S-module M.*

PROOF. (1) See [**52**, Theorem 7.8].

(2) Let $\operatorname{Hom}_S(Y,-) \to \operatorname{Hom}_S(X,-) \to F \to 0$ be a presentation of $F = \operatorname{Hom}_R(R,-) \circ f$ and choose an epimorphism $S^c \to X$ in $\operatorname{mod} S$. Then $c \cdot \ell_{\text{end}}(M)$ bounds the length of the $\operatorname{End}_S(M)^{\text{op}}$-module $F(M)$ and therefore also the endolength of $f(M)$ since the $\operatorname{End}_S(M)^{\text{op}}$-module structure on $F(M)$ is induced by that of $\operatorname{End}_R(f(M))^{\text{op}}$ via the canonical homomorphism $\operatorname{End}_S(M) \to \operatorname{End}_R(f(M))$.

(3) We use the adjoint pair of functors f_* and f^* between $D(R)$ and $D(S)$ which has been constructed in the proof of Theorem 9.1. The canonical ring homomorphism $\phi\colon \operatorname{End}_S(M) \to \operatorname{End}_R(f(M))$ induces the following commutative diagram of exact functors:

$$\begin{array}{ccc} C(R) & \xrightarrow{f^*} & C(S) \\ \downarrow{\scriptstyle E_{f(M)}} & & \downarrow{\scriptstyle E_M} \\ \operatorname{Mod}\operatorname{End}_R(f(M))^{\text{op}} & \xrightarrow{\phi_*} & \operatorname{Mod}\operatorname{End}_S(M)^{\text{op}} \end{array}$$

The assertion is now an immediate consequence of the definition of the endocategory of M and $f(M)$.

(4) Combining (3) with Proposition 7.6 and Lemma B.1 we obtain

$$\operatorname{KGdim} f(M) = \dim \mathcal{E}_{f(M)} \leq \dim \mathcal{E}_M = \operatorname{KGdim} M.$$

□

The preceding result motivates the following definition. Given a subset $\mathbf{U} \subseteq \operatorname{Ind} S$, we call a map $f\colon \mathbf{U} \to \operatorname{Mod} R$ *finite* provided that the following holds:

(F1) If $\mathbf{V} \subseteq \mathbf{U}$ is a Ziegler-closed subset of $\operatorname{Ind} S$, then the indecomposable direct summands of modules in $f(\mathbf{V})$ form a Ziegler-closed subset of $\operatorname{Ind} R$.

(F2) There is $c \in \mathbb{N}$ such that $\ell_{\text{end}}(f(M)) \leq c \cdot \ell_{\text{end}}(M)$ for all $M \in \mathbf{U}$.

COROLLARY 9.8. *A coherent functor* $\operatorname{Mod} S \to \operatorname{Mod} R$ *induces a finite map* $\operatorname{Ind} S \to \operatorname{Mod} R$.

The next result expresses the fact that coherent functors preserve the Ziegler topology. This generalizes the main result in [**62**] where so-called representation embeddings are discussed.

COROLLARY 9.9. *A coherent functor* $\operatorname{Mod} S \to \operatorname{Mod} R$ *which sends pure-injective indecomposables to indecomposables, induces a continuous and closed map* $\operatorname{Ind} S \to \operatorname{Ind} R$ *between the Ziegler spectra of S and R.*

We include another consequence of Theorem 9.1 which will be needed later. Suppose that R and S are k-algebras over some commutative noetherian ring k.

COROLLARY 9.10. *Let $f\colon \operatorname{Mod} S \to \operatorname{Mod} R$ be a coherent k-linear functor. Then $f(M)$ is finitely generated over k for every S-module M which is finitely generated over k.*

PROOF. $f(M)$ is a subquotient of M^n for some $n \in \mathbb{N}$. □

Our next aim is to show that a coherent functor $\operatorname{Mod} S \to \operatorname{Mod} R$ induces a homomorphisms between the K-groups of $\operatorname{mod} R$ and $\operatorname{mod} S$. We recall briefly the relevant concepts. In [**67**], Quillen constructs for each exact category $\mathcal{C}$ and each integer $i \geq 0$ an abelian group $K_i\mathcal{C}$. We shall use these groups for two types of exact categories. If $\mathcal{C}$ is an abelian category, then we denote by $K_i\mathcal{C}$ the K-groups with respect to the class of all exact sequences. If $\mathcal{C}$ is an additive category, then we denote by $K_i(\mathcal{C}, 0)$ the K-groups with respect to the class of all split exact sequences.

LEMMA 9.11. *The functor* $\operatorname{mod} R \to C(R)$, $X \mapsto T_X$, *induces an isomorphism* $K_i(\operatorname{mod} R, 0) \to K_i C(R)$ *for all* $i \geq 0$.

PROOF. The functor identifies $\operatorname{mod} R$ with the full subcategory of injective objects in $C(R)$. Every object F in $C(R)$ has a finite injective copresentation

$$0 \longrightarrow F \longrightarrow T_X \longrightarrow T_Y \longrightarrow T_Z \longrightarrow 0$$

The assertion now follows from a result in [**67**]. □

COROLLARY 9.12. *A coherent functor* $\operatorname{Mod} S \to \operatorname{Mod} R$ *induces a homomorphism* $K_i(\operatorname{mod} R, 0) \to K_i(\operatorname{mod} S, 0)$ *for all* $i \geq 0$.

PROOF. The homomorphism $K_i(\operatorname{mod} R, 0) \to K_i(\operatorname{mod} S, 0)$ is induced by the exact functor $C(R) \to C(S)$. □

REMARK 9.13. Let $f\colon \operatorname{Mod} S \to \operatorname{Mod} R$ be a coherent functor which is left exact (e.g. f is the restriction functor corresponding to a homomorphism $R \to S$). Then f has a left adjoint which restricts to a functor $\operatorname{mod} R \to \operatorname{mod} S$ by Lemma 1.1. It induces a homomorphism $K_i(\operatorname{mod} R, 0) \to K_i(\operatorname{mod} S, 0)$ for all $i \geq 0$ which coincides with the homomorpism constructed in the preceding corollary.

CHAPTER 10

Tame algebras

Suppose that R is a finite dimensional algebra over some algebraically closed field k. A *one-parameter family* of R-modules of dimension n is the set of R-modules

$$\{k[T]/(T-\lambda)\otimes_{k[T]} B \mid \lambda \in k\}$$

where B is a $k[T]$-R-bimodule which is free of rank n over $k[T]$. The algebra is said to be of *tame representation type* provided that there is for every $n \in \mathbb{N}$ a finite number of such one-parameter families such that every indecomposable R-module of dimension n is isomorphic to a module in one of these families [**24, 14**].

The main aim of this chapter is to present two new definitions of tame representation type. Both definitions seem to be more natural. The first one is formulated in terms of endofinite modules and behaves well with respect to functors between module categories. The second definition uses fp-idempotent ideals and is therefore entirely formulated in terms of the category of finitely presented modules.

10.1. Endofinitely tame algebras

The notion of tameness for finite dimensional algebras depends on the concept of a one-parameter family of finite dimensional modules. An alternative approach was developed by Crawley-Boevey. He used certain endofinite modules, so-called generic modules, and proved that a finite dimensional algebra over some algebraically closed field has tame representation type if and only if it is generically tame [**17**]. In this section we present a slight variation of Crawley-Boevey's concept.

We begin with some definitions. A module is called *generic* if it is endofinite indecomposable but not finitely presented [**17, 51**]. A noetherian algebra is *generically tame* provided that for every $n \in \mathbb{N}$ there are only finitely many generic modules of endolength n, see [**17**]. Recall that R is a *noetherian algebra* if the centre $Z(R)$ is noetherian and R is a finitely generated module over $Z(R)$. Given $n \in \mathbb{N}$, we denote by $\operatorname{ind}_n R$ the set of modules in $\operatorname{Ind} R$ which are finitely presented and have endolength n. We call a noetherian algebra *endofinitely tame* provided that for every $n \in \mathbb{N}$ the Ziegler closure of $\operatorname{ind}_n R$ contains only finitely many modules which are not finitely presented. Let us mention some basic properties of generically and endofinitely tame algebras.

PROPOSITION 10.1. *A noetherian algebra R is generically (endofinitely) tame if and only if R^{op} is generically (endofinitely) tame.*

PROOF. Apply the bijection $M \mapsto M^{\vee}$ between the endofinite modules over R and R^{op} from Corollary 6.19. Note that this bijection sends generic modules to generic modules; this follows from Proposition 6.23. Moreover, one uses that the map $M \mapsto M^{\vee}$ is compatible with the Ziegler topology by Proposition 4.17. □

PROPOSITION 10.2. *A generically tame noetherian algebra is endofinitely tame.*

PROOF. Every module in the Ziegler closure of $\operatorname{ind}_n R$ is endofinite of endolength at most n by Proposition 6.20. □

The following results give characterizations of generically and endofinitely tame algebras which are entirely formulated in terms of the category of finitely presented modules.

THEOREM 10.3. *An artin algebra R is generically tame if and only if for every $n \in \mathbb{N}$ there are only finitely many fp-idempotent and nilpotent ideals of length n in* $\operatorname{mod} R$.

PROOF. We use the bijection between finite sets of generic modules and certain fp-idempotent ideals. More precisely, it follows from Theorem 6.28 and Corollary 6.30 that an ideal $\mathfrak{I}$ is fp-idempotent and nilpotent of length n if and only if $\mathfrak{I} = [\mathbf{U}]$ for some subset $\mathbf{U} \subseteq \operatorname{Ind} R$ consisting of generic modules with $\sum_{M \in \mathbf{U}} \ell_{\text{end}}(M) = n$. □

A noetherian algebra R is called *domestic* if there are, up to isomorphism, only finitely many generic R-modules.

COROLLARY 10.4. *An artin algebra R is domestic if and only if there exists $n \in \mathbb{N}$ such that every fp-idempotent and nilpotent ideal in* $\operatorname{mod} R$ *has length at most n.*

We give now a characterization of endofinitely tame algebras.

THEOREM 10.5. *Let R be an artin algebra and denote for every $n \in \mathbb{N}$ by* $\operatorname{fpnil}_n R$ *the set of non-zero fp-idempotent and nilpotent ideals of* $\operatorname{mod} R$ *which are contained in* $[\operatorname{ind}_n R]$. *Then the map*

$$\overline{\operatorname{ind}_n R} \setminus \operatorname{ind}_n R \longrightarrow \operatorname{fpnil}_n R, \quad M \mapsto [M]$$

induces a bijection between the generic modules in the Ziegler closure of $\operatorname{ind}_n R$ *and the minimal elements of* $\operatorname{fpnil}_n R$. *Moreover, R is endofinitely tame if and only if* $\operatorname{fpnil}_n R$ *is finite for all $n \in \mathbb{N}$.*

PROOF. It follows from Corollary 5.14 and Corollary 5.16 that the Ziegler-closed subsets of $\overline{\operatorname{ind}_n R}$ correspond bijectively to the fp-idempotent ideals contained in $[\operatorname{ind}_n R]$. Every module M in $\overline{\operatorname{ind}_n R}$ is endofinite and therefore $[M]$ is the ideal corresponding to $\{M\}$ by Theorem 6.28. The assertion now follows since M is generic if and only if $[M]$ is nilpotent. □

10.2. Functors preserving tameness

In this section we show that coherent functors between module categories preserve tameness if certain density conditions are satisfied. This work is motivated by the fact that the original definition of a one-parameter family of finite dimensional modules over a k-algebra R amounts to a coherent functor $\operatorname{Mod} k[T] \to \operatorname{Mod} R$ which is k-linear and exact.

The first result of this section shows that endofinite tameness is well behaved with respect to functors between module categories. This follows from the fact that every coherent functor $\operatorname{Mod} S \to \operatorname{Mod} R$ induces a finite map $\operatorname{Ind} S \to \operatorname{Mod} R$. We refer to **[60]** for a discussion of similar results. Note that de la Peña uses a geometric approach; however the precise relation with the approach based on generic modules is still not clear.

PROPOSITION 10.6. *A noetherian algebra R is endofinitely tame if and only if for every $n \in \mathbb{N}$ there is an endofinitely tame noetherian algebra S, a subset $\mathbf{V} \subseteq \bigcup_{i=1}^{m} \operatorname{ind}_i S$ for some $m \in \mathbb{N}$, a finite map $f\colon \overline{\mathbf{V}} \to \operatorname{Mod} R$, and some cofinite subset $\mathbf{U} \subseteq \operatorname{ind}_n R$ such that $\mathbf{U} \subseteq \operatorname{add} f(\mathbf{V}) \subseteq \operatorname{mod} R$.*

PROOF. We fix $n \in \mathbb{N}$ and denote by $\overline{\mathbf{V}}$ the Ziegler closure of $\mathbf{V}$. By assumption there are only finitely many generic modules in $\overline{\mathbf{V}}$, and we claim that also the number of generic modules in $\operatorname{add} f(\overline{\mathbf{V}})$ is finite. In fact each $M \in \overline{\mathbf{V}}$ is endofinite by Proposition 6.20, and therefore $f(M)$ is endofinite since f is finite. Note also that $\overline{\mathbf{V}} \setminus \mathbf{V}$ is finite since there are no finitely presented modules in $\overline{\mathbf{V}} \setminus \mathbf{V}$ by Proposition 6.22. An endofinite module has only finitely many non-isomorphic indecomposable direct summands, for instance by Theorem 6.19, and therefore our first claim follows. Now suppose that $\operatorname{ind}_n R = \mathbf{U} \cup \mathbf{U}'$ for some finite set $\mathbf{U}'$, and observe that $\overline{\mathbf{U}} \cup \mathbf{U}'$ is the Ziegler closure of $\operatorname{ind}_n R$ since $\mathbf{U}'$ is Ziegler-closed by Proposition 6.17. Using again that f is finite, we have $\overline{\mathbf{U}} \subseteq \operatorname{add} f(\overline{\mathbf{V}})$ since $\operatorname{add} f(\overline{\mathbf{V}}) \cap \operatorname{Ind} R$ is a Ziegler-closed subset which contains $\mathbf{U}$ by assumption. Thus the Ziegler closure of $\operatorname{ind}_n R$ contains only finitely many modules which are not finitely presented, and therefore R is endofinitely tame. □

REMARK 10.7. *The requirement $f(\mathbf{V}) \subseteq \operatorname{mod} R$ is not automatically satisfied if the finite map $\overline{\mathbf{V}} \to \operatorname{Mod} R$ is induced by a coherent functor $\operatorname{Mod} S \to \operatorname{Mod} R$. This was pointed out by M. Prest. Take, for instance, the forgetful functor $\operatorname{Mod} \mathbb{Q} \to \operatorname{Mod} \mathbb{Z}$.*

We obtain the following consequence of the preceding proposition.

COROLLARY 10.8. *A noetherian algebra R is endofinitely tame if and only if for every $n \in \mathbb{N}$ there is an endofinitely tame noetherian algebra S, some $m \in \mathbb{N}$, and a coherent functor $f\colon \operatorname{Mod} S \to \operatorname{Mod} R$ such that all but finitely many modules in $\operatorname{ind}_n R$ occur as a direct summand of some finitely presented module $f(M)$ with $M \in \bigcup_{i=1}^{m} \operatorname{ind}_i S$.*

PROOF. Combine Proposition 10.6 with Corollary 9.8. □

From now on we assume that R is a finite dimensional algebra over some algebraically closed field k. We want to show that R is endofinitely tame if R is of tame representation type in the usual sense. To this end we need a reformulation of the classical definition.

LEMMA 10.9. *The algebra R is of tame representation type if and only if for every $n \in \mathbb{N}$ there is a coherent k-linear and exact functor $f\colon \operatorname{Mod} S \to \operatorname{Mod} R$ for some finite product $S = k[T] \times \ldots \times k[T]$, such that $\operatorname{ind}_n R \subseteq f(\operatorname{ind}_1 S)$.*

PROOF. Observe that endolength and k-dimension coincide for every finite dimensional indecomposable module over a k-algebra since k is algebraically closed. Using this fact the assertion follows from the next lemma. □

LEMMA 10.10. *For a functor $f\colon \operatorname{Mod} k[T] \to \operatorname{Mod} R$ the following are equivalent:*

(1) *f is coherent, k-linear, and exact.*
(2) *There is a $k[T]$-R-bimodule B on which k acts centrally and which is free of finite rank over $k[T]$, such that $f \simeq - \otimes_{k[T]} B$.*

PROOF. Left to the reader. □

THEOREM 10.11. *Let R be a finite dimensional algebra over some algebraically closed field. If R is of tame representation type, then R is endofinitely tame.*

PROOF. The unique generic module over $k[T]$ is the quotient field $k(T)$, and therefore any finite product $k[T] \times \ldots \times k[T]$ is generically tame. Using the reformulation of the usual tameness definition in Lemma 10.9, the assertion is an immediate consequence of Proposition 10.6 and Corollary 9.8. $\square$

10.3. Representation embeddings

We call a functor $f\colon \operatorname{Mod} S \to \operatorname{Mod} R$ a *representation embedding* if any two indecomposable endofinite S-modules M and N are isomorphic if and only if $\operatorname{add} f(M) \cap \operatorname{add} f(N) \neq 0$.

THEOREM 10.12. *Let R be a noetherian algebra over some infinite field k, and suppose there is a coherent k-linear representation embedding $f\colon \operatorname{Mod} k\langle X, Y\rangle \to \operatorname{Mod} R$. Then R is not endofinitely tame.*

PROOF. Consider for every pair $\alpha, \beta \in k$ the $k\langle X, Y\rangle$-module

$$M_{\alpha,\beta} = k\langle X, Y\rangle/(X - \alpha, Y - \beta)$$

and let G_α be the endofinite $k\langle X, Y\rangle$-module whose underlying space is $k(T)$ with X acting by multiplication with α and Y acting by multiplication with T. It is not hard to see that for each $\alpha \in k$ the Ziegler closure of $\mathbf{U}_\alpha = \{M_{\alpha,\beta} \mid \beta \in k\}$ is $\mathbf{V}_\alpha = \mathbf{U}_\alpha \cup \{G_\alpha\}$, for instance by arguments given in [**51**]. Denote by $\mathbf{U}'_\alpha$ the set of indecomposable direct summands of modules in $f(\mathbf{U}_\alpha)$. Analogously, $\mathbf{V}'_\alpha$ is defined which is a Ziegler-closed subset of $\operatorname{Ind} R$ by Corollary 9.8 since $\mathbf{V}_\alpha$ is Ziegler-closed. Also, there is $n \in \mathbb{N}$ such that each module in $\mathbf{V}'_\alpha$ has endolength at most n since every module in $\mathbf{V}_\alpha$ has endolength 1. Therefore $\mathbf{U}'_\alpha \subseteq \operatorname{ind}_n R$ for all α since $f(M)$ is finitely generated over k for all M in $\mathbf{U}_\alpha$ by Corollary 9.9. We claim that the Ziegler closure of $\operatorname{ind}_n R$ contains infinitely many generic modules. Using the fact that f is a representation embedding, it follows that $\operatorname{card} \mathbf{U}'_\alpha \geq \operatorname{card} \mathbf{U}_\alpha = \operatorname{card} k$. Thus the Ziegler closure of each $\mathbf{U}'_\alpha$ contains a generic module by Proposition 6.22. Also, $\mathbf{V}'_\alpha \cap \mathbf{V}'_\beta = \emptyset$ for $\alpha \neq \beta$ since $\mathbf{V}_\alpha \cap \mathbf{V}_\beta = \emptyset$ and f is a representation embedding. The assertion now follows since each $\mathbf{V}'_\alpha$ contains the Ziegler closure of $\mathbf{U}'_\alpha$. $\square$

From now on assume for the rest of this section that R is a finite dimensional algebra over some algebraically closed field. If R is of tame representation type, then we denote for every $n \in \mathbb{N}$ by $\mu_R(n)$ the minimal number of one-parameter families which is needed to parametrize, up to isomorphism, all but finitely many indecomposable R-modules of dimension n.

THEOREM 10.13. *The algebra R is of tame representation type if and only if it is endofinitely tame. Moreover, in this case $\mu_R(n) = \operatorname{card}(\overline{\operatorname{ind}_n R} \setminus \operatorname{ind}_n R)$ for every $n \in \mathbb{N}$.*

PROOF. It has already been shown in Theorem 10.11 that a tame algebra is endofinitely tame. Now suppose that R is not of tame representation type. Applying the Tame and Wild Theorem [**24, 17**], there exists a representation embedding $\operatorname{Mod}\langle X, Y\rangle \to \operatorname{Mod} R$. It follows from Theorem 10.12 that R is not endofinitely tame.

It remains to verify the assertion about $\mu_R(n)$. It has been shown in [**51**, Corollary 9.7] that a generic module M belongs to $\overline{\operatorname{ind}_n R}$ if and only if the endolength of M divides n. There are precisely $\mu_R(n)$ generic modules with this property by [**17**, Theorem 5.6], and we obtain therefore $\mu_R(n) = \operatorname{card}(\overline{\operatorname{ind}_n R} \setminus \operatorname{ind}_n R)$. $\square$

We are now in a position to give a characterization of tame algebras which is entirely formulated in terms of the category of finitely presented modules.

COROLLARY 10.14. *The algebra R is of tame representation type if and only if for every $n \in \mathbb{N}$ there are only finitely many non-zero fp-idempotent and nilpotent ideals in* $\operatorname{mod} R$ *which are contained in* $[\operatorname{ind}_n R]$. *Moreover, in this case $\mu_R(n)$ equals the number of minimal elements in the set of these ideals.*

10.4. One-parameter families and generic modules

In this section we describe explicitly the relation between one-parameter families and generic modules over algebras of tame representation type. The results presented here are mainly due to Crawley-Boevey; those results which involve the Ziegler spectrum are due to Krause. We do not give proofs but refer to Section 5 in [**17**] and Section 9 in [**51**]. Throughout this section R is a finite dimensional algebra over some algebraically closed field.

THEOREM 10.15. *Let R be of tame representation type and let $n \in \mathbb{N}$. Then there are one-parameter families $\mathbf{U}_1, \dots, \mathbf{U}_{\mu_R(n)}$ of n-dimensional indecomposable R-modules having the following properties:*

1. $\operatorname{ind}_n R \setminus (\mathbf{U}_1 \cup \dots \cup \mathbf{U}_{\mu_R(n)})$ *is finite.*
2. *The Ziegler closure $\overline{\mathbf{U}_i}$ contains a unique generic R-module G_i for every i.*
3. $G_i = G_j$ *if and only if* $i = j$.
4. $\overline{\operatorname{ind}_n R} \setminus \operatorname{ind}_n R = \{G_1, \dots, G_{\mu_R(n)}\}$.
5. *A generic R-module belongs to $\{G_1, \dots, G_{\mu_R(n)}\}$ if and only if its endolength divides n.*

In [**14**], Crawley-Boevey has shown that over a tame algebra almost all indecomposable modules of a fixed dimension belong to homogeneous tubes of the Auslander-Reiten quiver of R. Recall that a family $\mathcal{T} = (M_i)_{i \in \mathbb{N}}$ of R-modules forms a *homogeneous tube* if there are maps $M_i \to M_{i+1}$ and $M_{i+1} \to M_i$ for every $i \in \mathbb{N}$ which induce almost split sequences $0 \to M_i \to M_{i-1} \coprod M_{i+1} \to M_i \to 0$ for every $i \in \mathbb{N}$ where $M_0 = 0$. We say that $\mathcal{T}$ is *generic* if the Ziegler closure of the M_i is of the form $\mathcal{T} \cup \{\varinjlim M_i, \varprojlim M_i, G\}$ for some generic module G.

THEOREM 10.16. *Let R be of tame representation type and let $n \in \mathbb{N}$. Then all but finitely many n-dimensional indecomposable R-modules belong to generic homogeneous tubes.*

The occurence of homogeneous tubes is a characteristic phenomenon for tame algebras. This can be made precise as follows.

THEOREM 10.17. *An algebra R is of tame representation type if and only if every generic R-module belongs to the Ziegler closure of a (generic) homogeneous tube.*

CHAPTER 11

Rings of definable scalars

In this chapter we assign to every collection Φ of maps in $\operatorname{mod} R$ a ring homomorphism $f_\Phi\colon R \to R_\Phi$ which is called the ring of definable scalars for Φ. The homomorphism $R \to R_\Phi$ depends only on the saturation $\overline{\Phi}$ of Φ and is therefore an invariant of the definable subcategory $(\operatorname{Mod} R)_\Phi$. We use a universal property to define the ring of definable scalars but present various alternative constructions for $R \to R_\Phi$.

11.1. Rings of definable scalars

The following existence theorem describes the ring of definable scalars as an invariant of a definable subcategory.

THEOREM 11.1. *Let $\mathcal{X}$ be a definable subcategory of* $\operatorname{Mod} R$. *Then there exists a ring homomorphism $f\colon R \to S$ and a definable subcategory $\mathcal{Y}$ of* $\operatorname{Mod} S$ *satisfying the following properties:*

(1) *Restriction via f induces an equivalence $\mathcal{Y} \to \mathcal{X}$.*
(2) *Suppose there is a ring homomorphism $f'\colon R \to S'$ and a definable subcategory $\mathcal{Y}'$ of* $\operatorname{Mod} S'$ *such that restriction via f' induces an equivalence $\mathcal{Y}' \to \mathcal{X}$. Then there is a unique ring homomorphism $g\colon S' \to S$ such that $f = g \circ f'$ and restriction via g induces an equivalence $\mathcal{Y} \to \mathcal{Y}'$.*

Moreover, the pair $(f, \mathcal{Y})$ is unique up to isomorphism.

Given a collection Φ of maps in $\operatorname{mod} R$, we denote by $f_\Phi\colon R \to R_\Phi$ the ring homomorphism $R \to S$ satisfying the conditions (1) – (2) in the preceding theorem for $\mathcal{X} = (\operatorname{Mod} R)_\Phi$, and we denote by Φ° the saturated collection of maps in $\operatorname{mod} R_\Phi$ such that $\mathcal{Y} = (\operatorname{Mod} R_\Phi)_{\Phi^\circ}$. Following Prest, we call the ring homomorphism $f_\Phi\colon R \to R_\Phi$ the *ring of definable scalars* for Φ. In [**63**], Prest introduced the ring of definable scalars for a closed subset of the Ziegler spectrum of R in model-theoretic terms but it can be shown that both concepts coincide. In fact, it is often useful to consider the ring of definable scalars for a class $\mathcal{C}$ of R-modules. We define $f_{\mathcal{C}}\colon R \to R_{\mathcal{C}}$ to be the ring of definable scalars for $\Phi_{\mathcal{C}}$ where $\Phi_{\mathcal{C}}$ denotes the collection of maps ϕ in $\operatorname{mod} R$ such that $\operatorname{Coker} H_\phi(M) = 0$ for all M in $\mathcal{C}$.

We begin our discussion with some notation. Let $f\colon R \to S$ be a ring homomorphism. The assignment $T_M \mapsto T_{M \otimes_R S}$ for any R-module M induces an exact functor $f^*\colon D(R) \to D(S)$. There is a right adjoint $f_*\colon D(S) \to D(R)$ which sends for every S-module M the functor T_M to T_{M_f} where M_f denotes the restriction of M via f.

LEMMA 11.2. *Let $\Phi \subseteq \operatorname{mod} R$ and $\Psi \subseteq \operatorname{mod} S$. Then the following are equivalent:*

(1) *The restriction of every Ψ-injective S-module is Φ-injective.*

(2) *$f^*(F) \in \mathcal{S}_\Psi$ for all $F \in \mathcal{S}_\Phi$.*
(3) *There is a (unique and exact) functor $f^*_{\Phi\Psi}$ making the following diagram of functors commutative.*

$$\begin{array}{ccc} C(R) & \xrightarrow{p_\Phi} & C(R)/\mathcal{S}_\Phi \\ \downarrow{\scriptstyle f^*} & & \downarrow{\scriptstyle f^*_{\Phi\Psi}} \\ C(S) & \xrightarrow{p_\Psi} & C(S)/\mathcal{S}_\Psi \end{array}$$

Moreover, restriction via f induces an equivalence $(\operatorname{Mod} S)_\Psi \to (\operatorname{Mod} R)_\Phi$ if and only if $f^*_{\Phi\Psi}$ is an equivalence.

PROOF. (1) $\Leftrightarrow$ (2) It has been shown in Theorem 2.1 that an R-module M is Φ-injective if and only if $\operatorname{Hom}(\mathcal{S}_\Phi, T_M) = 0$. Using this fact the assertion follows from the adjointness isomorphism $\operatorname{Hom}(f^*(F), T_M) \simeq \operatorname{Hom}(F, T_{M_f})$ where M is any S-module and M_f denotes its restriction via f.

(2) $\Leftrightarrow$ (3) Condition (2) holds if and only if $p_\Psi \circ f^*(\mathcal{S}_\Phi) = 0$ and therefore the universal property of p_Φ implies the existence of $f^*_{\Phi\Psi}$ whenever (2) holds.

The last statement follows with Corollary 2.4 where it is shown that $(\operatorname{Mod} R)_\Phi$ and the category of exact functors $(C(R)/\mathcal{S}_\Phi)^{\mathrm{op}} \to \mathrm{Ab}$ are canonically equivalent. □

We shall need the following technical lemma.

LEMMA 11.3. *Let $\mathcal{C} \xrightarrow{f} \mathcal{D} \xrightarrow{g} \mathcal{C}/\mathcal{S}$ be a sequence of exact functors between abelian cateories, and suppose that the composition is the quotient functor corresponding to the Serre subcategory $\mathcal{S}$. If $\mathcal{T} = \operatorname{Ker} g$, then g induces an equivalence $\mathcal{D}/\mathcal{T} \to \mathcal{C}/\mathcal{S}$.*

PROOF. Let $q\colon \mathcal{D} \to \mathcal{D}/\mathcal{T}$ denote the quotient functor corresponding to $\mathcal{T}$. Clearly, the induced functor $h\colon \mathcal{D}/\mathcal{T} \to \mathcal{C}/\mathcal{S}$ is faithful and dense. To see that h is full let $\alpha\colon X \to Y$ be a morphism in $\mathcal{C}/\mathcal{S}$. By definition, there are subobjects $X' \subseteq X$ and $Y' \subseteq Y$ such that $g \circ f(\beta) \simeq \alpha$ for some morphism $\beta\colon X' \to Y/Y'$ in $\mathcal{C}$. Composing $q \circ f(\beta)$ with inverses of the inclusion $q \circ f(X') \to q \circ f(X)$ and the projection $q \circ f(Y) \to q \circ f(Y/Y')$ we obtain a morphism γ with $h(\gamma) = \alpha$. □

PROOF OF THEOREM 11.1. Let $\mathcal{X} = (\operatorname{Mod} R)_\Phi$ be a definable subcategory. We consider the quotient functor $p_\Phi\colon C(R) \to C(R)/\mathcal{S}_\Phi$. Let $S = \operatorname{End}_{C(R)/\mathcal{S}_\Phi}(T_R)$ and let $f\colon R \to S$ be the homomorphism induced by p_Φ. The homomorphism f extends to an exact functor $f^*\colon C(R) \to C(S)$ and $\mathrm{id}\colon S \to \operatorname{End}_{C(R)/\mathcal{S}_\Phi}(T_R)$ extends to an exact functor $\mathrm{id}'\colon C(S) \to C(R)/\mathcal{S}_\Phi$ which satisfy $p_\Phi = \mathrm{id}' \circ f^*$ by Lemma 1.2. We define Ψ to be the collection of maps ϕ in $\operatorname{mod} S$ such that $\mathrm{id}'(\operatorname{Ker} T_\phi) = 0$ and let $\mathcal{Y} = (\operatorname{Mod} S)_\Psi$. It follows that $f^*(F) \in \mathcal{S}_\Psi$ for every $F \in \mathcal{S}_\Phi$, and therefore the restriction via f of every Ψ-injective S-module is Φ-injective by Lemma 11.2. In fact, the induced functor $\mathcal{Y} \to \mathcal{X}$ is an equivalence since $\mathrm{id}'\colon C(S) \to C(R)/\mathcal{S}_\Phi$ induces an equivalence $C(S)/\mathcal{S}_\Psi \to C(R)/\mathcal{S}_\Phi$ by Lemma 11.3 which is an inverse of $f^*_{\Phi\Psi}$. Having shown condition (1) it remains to verify (2). To this end suppose there is a ring homomorphism $g\colon R \to T$ and a definable subcategory $\mathcal{Z} = (\operatorname{Mod} T)_\Omega$ such that restriction via g induces an equivalence $\mathcal{Z} \to \mathcal{X}$. It follows from Lemma 11.2 that g^* induces an equivalence $g^*_{\Phi\Omega}\colon C(R)/\mathcal{S}_\Phi \to C(T)/\mathcal{S}_\Omega$. Composing the homomorphism $T \to \operatorname{End}_{C(T)/\mathcal{S}_\Omega}(T_T)$ induced by p_Ω with the inverse of the isomorphism $S \to \operatorname{End}_{C(T)/\mathcal{S}_\Omega}(T_T)$ induced by $g^*_{\Phi\Omega}$ we obtain a homomorphism $h\colon T \to S$ with

$f = h \circ g$. The following commutative diagram of exact functors illustrates the construction.

$$\begin{array}{ccccc} C(R) & \xrightarrow{g^*} & C(T) & \xrightarrow{h^*} & C(S) \\ \downarrow{\scriptstyle p_\Phi} & & \downarrow{\scriptstyle p_\Omega} & & \downarrow{\scriptstyle p_\Psi} \\ C(R)/\mathcal{S}_\Phi & \xrightarrow{g^*_{\Phi\Omega}} & C(T)/\mathcal{S}_\Omega & \xrightarrow{h^*_{\Omega\Psi}} & C(S)/\mathcal{S}_\Psi \end{array}$$

Note that $h^*_{\Omega\Psi}$ is the composition of an inverse of $g^*_{\Phi\Omega}$ with $f^*_{\Phi\Psi}$. Therefore restriction via h induces an equivalence $\mathcal{Y} \to \mathcal{Z}$, and we leave it to the reader to deduce the uniqueness of the homomorphism h from the commutative diagram. □

Having shown the existence of the ring of definable scalars, we now continue with a discussion of basic properties and alternative descriptions. Our first statement is essentially a reformulation of the construction given in the preceding proof.

COROLLARY 11.4. *For a ring homomorphism $f\colon R \to S$ the following are equivalent:*

(1) *f is the ring of definable scalars for a collection of maps in* $\operatorname{mod} R$.
(2) *f is isomorphic to the ring homomorphism* $\operatorname{End}_{C(R)}(T_R) \to \operatorname{End}_{C(R)/\mathcal{S}}(T_R)$ *which is induced by the quotient functor $C(R) \to C(R)/\mathcal{S}$ for some Serre subcategory $\mathcal{S}$ of $C(R)$.*

It is interesting to note that Morita equivalent rings have Morita equivalent rings of definable scalars.

COROLLARY 11.5. *Let R and S be Morita equivalent rings and fix an equivalence $f\colon \operatorname{mod} R \to \operatorname{mod} S$. If Φ is a collection of maps in $\operatorname{mod} R$, then R_Φ and $S_{f(\Phi)}$ are Morita equivalent.*

PROOF. The equivalence f extends to an equivalence $f'\colon C(R) \to C(S)$ via $T_X \mapsto T_{f(X)}$. It is clear that f' induces an equivalence $C(R)/\mathcal{S}_\Phi \to C(S)/\mathcal{S}_{f(\Phi)}$. Therefore $R_\Phi = \operatorname{End}_{C(R)/\mathcal{S}_\Phi}(T_R)$ and $S_{f(\Phi)} = \operatorname{End}_{C(S)/\mathcal{S}_{f(\Phi)}}(T_S)$ are Morita equivalent. □

11.2. Calculus of fractions

Let Φ be a saturated collection of maps in $\operatorname{mod} R$. A *left fraction* with respect to Φ is a pair (α, ϕ) of maps $\alpha, \phi\colon R \to X$ in $\operatorname{mod} R$ such that ϕ and $c(c(\phi) \circ \alpha)$ belong to Φ. Here, we denote for any map $\phi\colon X \to Y$ by $c(\phi)$ the cokernel map $Y \to \operatorname{Coker} \phi$. The collection of left fractions with respect to Φ is denoted by $R(\Phi^-)$. Given left fractions (α_1, ϕ_1) and (α_2, ϕ_2) we consider the following pushout diagrams:

$$\begin{array}{ccc} R & \xrightarrow{\phi_2} & X_2 \\ \downarrow{\scriptstyle \phi_1} & & \downarrow{\scriptstyle \gamma_2} \\ X_1 & \xrightarrow{\gamma_1} & Y \end{array} \qquad \begin{array}{ccc} R & \xrightarrow{\alpha_2} & X_2 \\ \downarrow{\scriptstyle \phi_1} & & \downarrow{\scriptstyle \delta_2} \\ X_1 & \xrightarrow{\delta_1} & Z \end{array}$$

We obtain an equivalence relation on $R(\Phi^-)$ by defining $(\alpha_1, \phi_1) \sim (\alpha_2, \phi_2)$ if there are maps $\beta_i\colon X_i \to X$ such that $(\beta_1 \circ \alpha_1, \beta_1 \circ \phi_1)$ and $(\beta_2 \circ \alpha_2, \beta_2 \circ \phi_2)$ are equal and belong to $R(\Phi^-)$. The equivalence class of a left fraction (α, ϕ) is denoted by $[\alpha, \phi]$, and $R[\Phi^-]$ denotes the set of all equivalence classes in $R(\Phi^-)$. Addition and multiplication in $R[\Phi^-]$ are defined as follows:

$$[\alpha_1, \phi_1] + [\alpha_2, \phi_2] = [\gamma_1 \circ \alpha_1 + \gamma_2 \circ \alpha_2, \gamma_1 \circ \phi_1]$$

$$[\alpha_2, \phi_2] \circ [\alpha_1, \phi_1] = [\delta_1 \circ \alpha_1, \delta_2 \circ \phi_2]$$

Consider now the quotient functor $p\colon C(R) \to C(R)/\mathcal{S}_\Phi$ with respect to the Serre subcategory $\mathcal{S}_\Phi = \{\operatorname{Ker} T_\phi \mid \phi \in \Phi\}$ and recall that $R_\Phi = \operatorname{End}_{C(R)/\mathcal{S}_\Phi}(T_R)$. Given a left fraction (α, ϕ), the condition $\phi \in \Phi$ implies that $p(T_\phi)$ is a monomorphism, i.e. $p(T_\phi)$ is the kernel of $c(p(T_\phi))$. Furthermore, the condition $c(c(\phi) \circ \alpha) \in \Phi$ implies that $c(p(T_\phi)) \circ p(T_\alpha) = 0$ since the functor $X \mapsto p(T_X)$ on $\operatorname{mod} R$ is right exact. Therefore $p(T_\alpha)$ induces a unique morphism $\overline{\alpha}\colon T_R \to T_R$ such that $p(T_\alpha) = p(T_\phi) \circ \overline{\alpha}$.

THEOREM 11.6. *Let Φ be a collection of maps in* $\operatorname{mod} R$. *Then the set $R[\Phi^-]$ of equivalence classes of left fractions with respect to the saturation $\overline{\Phi}$ forms a ring. The assignment $[\alpha, \phi] \mapsto \overline{\alpha}$ induces an isomorphism between $R[\Phi^-]$ and the ring of definable scalars R_Φ.*

PROOF. Let $A = T_R = B$. By definition

$$R_\Phi = \operatorname{Hom}_{C(R)/\mathcal{S}_\Phi}(A, B) = \varinjlim \operatorname{Hom}_{C(R)}(A', B/B')$$

where $A/A', B' \in \mathcal{S}_\Phi$. Therefore every element in R_Φ is of the form $\pi(\rho)$ for some $\rho\colon A' \to B/B'$ where $\pi = (\pi_{A',B/B'})$ denotes the family of structural maps $\pi_{A',B/B'}\colon \operatorname{Hom}_{C(R)}(A', B/B') \to R_\Phi$. Choosing a monomorphism $\sigma\colon B/B' \to T_X$ for some X in $\operatorname{mod} R$, the composition $\sigma \circ \rho$ extends to a morphism $A \to T_X$ which is of the form T_α for some map $\alpha\colon R \to X$. Also, the composition of $B \to B/B'$ with σ is of the form T_ϕ for some map $\phi\colon R \to X$. It is easily checked that the pair (α, ϕ) is a left fraction with respect to $\overline{\Phi}$, and $\pi(\rho) = \overline{\alpha}$ holds by construction. Therefore the map $[\alpha, \phi] \mapsto \overline{\alpha}$ is surjective and it remains to check the injectivity. To this end suppose there are left fractions (α_1, ϕ_1) and (α_2, ϕ_2) with $\overline{\alpha_1} = \overline{\alpha_2}$. We claim that $(\alpha_1, \phi_1) \sim (\alpha_2, \phi_2)$. In fact, $p(T_\alpha) = 0$ for $\alpha = \gamma_1 \circ \alpha_1 - \gamma_2 \circ \alpha_2$ and therefore $\operatorname{Im} T_\alpha \in \mathcal{S}_\Phi$. Choosing a monomorphism $\sigma\colon T_Y/\operatorname{Im} T_\alpha \to T_X$ for some X in $\operatorname{mod} R$, the composition of $T_Y \to T_Y/\operatorname{Im} T_\alpha$ with σ is of the form T_β for some map $\beta\colon Y \to X$. We obtain

$$(\beta_1 \circ \alpha_1, \beta_1 \circ \phi_1) = (\beta_2 \circ \alpha_2, \beta_2 \circ \phi_2) \in R(\Phi^-)$$

for $\beta_i = \beta \circ \gamma_i$ and therefore $(\alpha_1, \phi_1) \sim (\alpha_2, \phi_2)$. □

11.3. Biendomorphism rings

Our next goal is the description of the ring of definable scalars as the biendomorphism ring of a certain module. This was done independently by Burke and Prest [**12**]; however our proof is substantially different. The approach presented here follows the exposition in the preprint version of [**52**]; it uses some localization theory for the functor category $D(R)$.

Let Φ be a collection of maps in $\operatorname{mod} R$. Let N_Φ be a product of a representative set of indecomposable pure-injective R-modules which are Φ-injective, and denote by M_Φ a product of κ copies of N_Φ where $\kappa = \operatorname{card} N_\Phi$. Given an R-module M with $S = \operatorname{End}_R(M)^{\mathrm{op}}$, the functor $\operatorname{Hom}_R(-, M)\colon \operatorname{Mod} R \to \operatorname{Mod} S$ sends R to M and induces a canonical homomorphism

$$R = \operatorname{End}_R(R) \longrightarrow \operatorname{End}_S(M)^{\mathrm{op}} = \operatorname{Biend}_R(M)$$

THEOREM 11.7. *The canonical homomorphism $R \to \operatorname{Biend}_R(M_\Phi)$ is the ring of definable scalars for Φ.*

Our proof depends on some general facts which we present now. Let $\mathcal{A}$ be an abelian category with products and suppose that $q\colon \mathcal{A} \to \mathcal{A}/\mathcal{T}$ is the quotient functor corresponding to a localizing subcategory $\mathcal{T}$. Let X and N be objects in $\mathcal{A}$ such that

(1) N is injective and $\operatorname{Hom}_{\mathcal{A}}(\mathcal{T}, N) = 0$;
(2) N cogenerates $\mathcal{A}/\mathcal{T}$;

and define $M = M^I$ with $S = \operatorname{End}_{\mathcal{A}}(M)^{\mathrm{op}}$ where $I = \operatorname{Hom}_{\mathcal{A}}(X, N)$.

LEMMA 11.8. *Denote by $f\colon \operatorname{End}_{\mathcal{A}}(X) \to \operatorname{End}_{\mathcal{A}/\mathcal{T}}(X)$ the homomorphism induced by the quotient functor $\mathcal{A} \to \mathcal{A}/\mathcal{T}$. Then*

$$h\colon \operatorname{End}_{\mathcal{A}}(X) \longrightarrow \operatorname{End}_S(\operatorname{Hom}_{\mathcal{A}}(X, M))^{\mathrm{op}}, \quad \phi \mapsto \operatorname{Hom}_{\mathcal{A}}(\phi, M)$$

induces an isomorphism $g\colon \operatorname{End}_{\mathcal{A}/\mathcal{T}}(X) \to \operatorname{End}_S(\operatorname{Hom}_{\mathcal{A}}(X, M))^{\mathrm{op}}$ such that $h = g \circ f$.

PROOF. The assumption $\operatorname{Hom}_{\mathcal{A}}(\mathcal{T}, M) = 0 = \operatorname{Ext}^1(\mathcal{T}, M)$ on M implies that q induces an isomorphism $\operatorname{Hom}_{\mathcal{A}}(Y, M) \to \operatorname{Hom}_{\mathcal{A}/\mathcal{T}}(Y, M)$ for all Y in $\mathcal{A}$ since M is $\mathcal{T}$-closed. Thus h factors through f. The following lemma then shows that the induced map g is an isomorphism because $\operatorname{Hom}_{\mathcal{A}}(X, M)$ is generated over S by $(\alpha_\phi)\colon X \to \prod_{\phi \in \operatorname{Hom}_{\mathcal{A}}(X,N)} N$ with $\alpha_\phi = \phi$ for all ϕ. □

LEMMA 11.9. *Let X and M be objects in any abelian category. If M is an injective cogenerator and* $\operatorname{Hom}(X, M)$ *is finitely generated as an $S = \operatorname{End}(M)^{\mathrm{op}}$-module, then* $\operatorname{Hom}(-, M)$ *induces an isomorphism* $\operatorname{End}(X) \to \operatorname{End}_S(\operatorname{Hom}(X, M))^{\mathrm{op}}$.

PROOF. The homomorphism is injective since M is a cogenerator. To show that it is surjective one uses in addition the assumption that $\operatorname{Hom}(X, M)$ is finitely generated over S so that X embeds into a finite product of copies of M. □

PROOF OF THEOREM 11.7. We apply Lemma 11.8 as follows. Let $\mathcal{A} = D(R)$, $\mathcal{T} = \mathcal{T}_\Phi$, $X = T_R$, and $N = T_{N_\phi}$ which implies $M = T_{M_\phi}$. The conditions (1) and (2) are immediate consequences of Proposition A.9. Using the fully faithful functor $\operatorname{Mod} R \to D(R)$ we obtain the following commutative diagram from Lemma 11.8

$$\begin{array}{ccccc}
R & \xrightarrow{f_\Phi} & R_\Phi & \longrightarrow & \operatorname{Biend}_R(M) \\
\downarrow \wr & & \| & & \downarrow \wr \\
\operatorname{End}_{\mathcal{A}}(X) & \xrightarrow{f} & \operatorname{End}_{\mathcal{A}/\mathcal{T}}(X) & \xrightarrow{g} & \operatorname{End}_S(\operatorname{Hom}_{\mathcal{A}}(X, M))^{\mathrm{op}}
\end{array}$$

which establishes the isomorphism between R_Φ and $\operatorname{Biend}_R(M_\Phi)$. □

An analysis of the preceding proof shows that in some cases the module M_Φ is far too big. We formulate therefore a variation of the preceding theorem which covers the case that Φ corresponds to a definable subcategory of the form $\operatorname{Add} M$ for some endofinite module M. To this end denote for an R-module M by Φ_M the collection of maps ϕ in $\operatorname{mod} R$ such that M is ϕ-injective. Clearly, the definable subcategory corresponding to Φ_M is the smallest containing M. Recall that a module M is Σ-pure-injective if every coproduct of copies of M is pure-injective. For example, every endofinite module is Σ-pure-injective.

COROLLARY 11.10. *Let M be a Σ-pure-injective R-module and suppose that M is finitely generated over* $\operatorname{End}_R(M)^{\mathrm{op}}$. *Then the canonical homomorphism $R \to$* $\operatorname{Biend}_R(M)$ *is the ring of definable scalars for Φ_M.*

PROOF. Adapt the argument of the proof of Theorem 11.7, and use the fact that T_M is an injective cogenerator for $D(R)/\mathcal{T}$ since M is Σ-pure-injective. □

EXAMPLE 11.11. A product-complete module is Σ-pure-injective and finitely generated over its endomorphism ring.

11.4. Basic properties

The kernel of the homomorphism $R \to R_\Phi$ can be described as follows.

PROPOSITION 11.12. *The kernel of $R \to R_\Phi$ is the annihilator of the class of Φ-injective R-modules.*

PROOF. Let $x \in R = \operatorname{End}_R(R)$. By definition of $f_\Phi \colon R \to R_\Phi$, we have $f_\Phi(x) = 0$ if and only if $\operatorname{Hom}(T_x, T_M) = 0$ for each Φ-injective R-module M. However, $\operatorname{Hom}(T_x, T_M) = 0$ if and only if $Mx = 0$. The assertion now follows. □

The next result reflects our functorial construction of the ring of definable scalars.

PROPOSITION 11.13. *Let Φ and Ψ be collections of maps in* $\operatorname{mod} R$ *and suppose that every Ψ-injective R-module is Φ-injective. Then there is a unique homomorphism $f_{\Phi\Psi} \colon R_\Phi \to R_\Psi$ such that $f_\Psi = f_{\Phi\Psi} \circ f_\Phi$ and $\phi \otimes_{R_\Phi} R_\Psi \in \Psi^\circ$ for all $\phi \in \Phi^\circ$.*

PROOF. The assumption on Φ and Ψ implies that $\operatorname{id} \colon R \to R$ induces an exact functor $\operatorname{id}^*_{\Phi\Psi} \colon C(R)/\mathcal{S}_\Phi \to C(R)/\mathcal{S}_\Psi$ in the sense of Lemma 11.2. We denote by $f_{\Phi\Psi} \colon R_\Phi \to R_\Psi$ the homomorphism which $\operatorname{id}^*_{\Phi\Psi}$ induces between the endomorphism rings of T_R. It is clear that $f_\Psi = f_{\Phi\Psi} \circ f_\Phi$ and the second condition is included to obtain the uniqueness of $f_{\Phi\Psi}$. □

COROLLARY 11.14. *Let $\Phi = \bigcup_{i \in I} \Phi_i$ be a directed union of collections of maps in* $\operatorname{mod} R$. *Then the homomorphisms $f_{\Phi_i\Phi} \colon R_{\Phi_i} \to R_\Phi$ induce an isomorphism $\varinjlim R_{\Phi_i} \to R_\Phi$.*

Given a collection Φ of maps in $\operatorname{mod} R$, we are now interested in ring homomorphisms $R \to S$ which factor through the ring of definable scalars R_Φ. We shall present a sufficient condition which in some sense is also necessary. The following lemma is needed.

LEMMA 11.15. *Let $f \colon R \to S$ be a ring homomorpism. Then the following are equivalent for a map ϕ in* $\operatorname{mod} R$*:*

(1) *The restriction of every S-module is ϕ-injective.*
(2) *$\phi \otimes_R S$ is a split mono in* $\operatorname{Mod} S$.
(3) $\operatorname{Ker} T_\phi$ *lies in the kernel of the exact functor $C(R) \to C(S)$ extending f.*

PROOF. (1) $\Leftrightarrow$ (2) This follows directly from the isomorphism $\operatorname{Hom}_R(X, M) \simeq \operatorname{Hom}_S(X \otimes_R S, M)$ for X in $\operatorname{Mod} R$ and M in $\operatorname{Mod} S$.

(2) $\Leftrightarrow$ (3) Let $\phi \otimes_R S = \psi \colon X \to Y$. The functor $C(R) \to C(S)$ sends T_ϕ to T_ψ and the assertion follows from the fact that T_ψ is a monomorphism if and only if T_ψ is a split monomorphism since T_X is fp-injective. □

THEOREM 11.16. *Let Φ be a collection of maps in* $\operatorname{mod} R$ *and $f \colon R \to S$ be a ring homomorphism. Then the following conditions are equivalent:*

(1) *The restriction of every S-module is Φ-injective.*
(2) *$\phi \otimes_R S$ is a split mono in* $\operatorname{Mod} S$ *for all $\phi \in \Phi$.*

(3) *There is a (unique) homomorphism* $g\colon R_\Phi \to S$ *such that* $f = g \circ f_\Phi$ *and the restriction via* g *of every* S*-module is* Φ°*-injective.*

PROOF. (1) $\Leftrightarrow$ (2) follows from the preceding lemma.

(1) $\Rightarrow$ (3) Lemma 11.2 implies that f^* induces a functor $f^*_{\Phi\emptyset}\colon C(R)/\mathcal{S}_\Phi \to C(S)$ which is exact. It induces a homomorphism $g\colon R_\Phi \to S$ having the required properties. The uniqueness of g is a consequence of the uniqueness of $f^*_{\Phi\emptyset}$ and the fact that g_* sends $\operatorname{Mod} S$ into $(\operatorname{Mod} R_\Phi)_{\Phi^\circ}$.

(3) $\Rightarrow$ (1) follows from the fact that $(f_\Phi)_*$ sends $(\operatorname{Mod} R_\Phi)_{\Phi^\circ}$ into $(\operatorname{Mod} R)_\Phi$. □

11.5. Epimorphisms of rings

It is well-known that every localization $R \to S$ of a ring R is an epimorphism in the category of rings. This section is devoted to a discussion of various converses of this statement. Our first result serves as the basis for all further results in this direction. It shows that every epimorphism $R \to S$ induces a localization functor $D(R) \to D(S)$.

PROPOSITION 11.17. *The following are equivalent for a ring homomorphism* $f\colon R \to S$*:*

(1) f *is an epimorphism.*
(2) *The exact functor* $f'\colon C(R) \to C(S)$ *extending* f *induces an equivalence* $C(R)/\mathcal{S} \to C(S)$ *for* $\mathcal{S} = \operatorname{Ker} f'$.
(3) *The exact functor* $f^*\colon D(R) \to D(S)$ *extending* $M \mapsto M \otimes_R S$ *induces an equivalence* $D(R)/\mathcal{T} \to D(S)$ *for* $\mathcal{T} = \operatorname{Ker} f^*$.

The proof uses the well-known fact that $f\colon R \to S$ is an epimorphism if and only if the restriction functor $\operatorname{Mod} S \to \operatorname{Mod} R$ is fully faithful. We shall also use the following lemma.

LEMMA 11.18. *Let* $q\colon \mathcal{A} \to \mathcal{B}$ *be an exact functor between two Grothendieck categories. Let* $\mathcal{C} = \operatorname{Ker} q$ *and suppose there exists a right adjoint* $s\colon \mathcal{B} \to \mathcal{A}$*. Then the following are equivalent.*

(1) q *induces an equivalence* $\mathcal{A}/\mathcal{C} \to \mathcal{B}$.
(2) s *is fully faithful.*
(3) s *induces a fully faithful functor* $\operatorname{inj} \mathcal{B} \to \operatorname{inj} \mathcal{A}$ *between the full subcategories of injective objects in* $\mathcal{B}$ *and* $\mathcal{A}$*, respectively.*

PROOF. See [**26**]. □

PROOF OF PROPOSITION 11.17. (1) $\Leftrightarrow$ (3) The functor $f^*\colon D(R) \to D(S)$ has a right adjoint f_* which extends the restriction functor $\operatorname{Mod} S \to \operatorname{Mod} R$. Note that f_* sends injective objects to injective objects since f^* is exact. However, the injective objects in $D(R)$ and $D(S)$ correspond under the functor $M \mapsto T_M$, up to isomorphism, precisely to the pure-injective modules. Therefore the assertion follows from the preceding lemma.

(2) $\Leftrightarrow$ (3) Use the fact that $\mathcal{S}$ and $\mathcal{T}$ are related via $\mathcal{T} = \varinjlim \mathcal{S}$ and $\mathcal{S} = \mathcal{T} \cap C(R)$. This follows from Proposition A.4. □

COROLLARY 11.19. *Let* $f\colon R \to S$ *be an epimorphism. Then the functor* $- \otimes_R S\colon \operatorname{mod} R \to \operatorname{mod} S$ *induces a long exact sequence*

$$\dots \longrightarrow K_1(\operatorname{mod} S, 0) \longrightarrow K_0 \operatorname{Ker} f' \longrightarrow K_0(\operatorname{mod} R, 0) \longrightarrow K_0(\operatorname{mod} S, 0) \longrightarrow 0$$

The kernel of $K_0(\operatorname{mod} R, 0) \longrightarrow K_0(\operatorname{mod} S, 0)$ *is generated by the elements* $[X] - [Y] + [\operatorname{Coker}\phi]$, *one for each map* $\phi\colon X \to X$ *in* $\operatorname{mod} R$ *such that* $\phi \otimes_R S$ *is a split mono.*

PROOF. Apply Lemma 9.11 and use the long exact sequence of K-groups [**67**] which is induced by the sequence $\operatorname{Ker} f' \to C(R) \xrightarrow{f'} C(S)$ where f' denotes the exact functor which extends f. The description of the kernel in degree 0 follows from Lemma 11.15. □

To state our next result, recall that a morphism $\alpha\colon A \to B$ is *left minimal* if every endomorphism $\beta\colon B \to B$ satisfying $\alpha = \beta \circ \alpha$ is an isomorphism.

THEOREM 11.20. *Let* $f\colon R \to S$ *be a ring homomorphism. Denote by* Φ *the collection of maps* ϕ *in* $\operatorname{mod} R$ *such that* $\phi \otimes_R S$ *is a split mono in* $\operatorname{mod} S$. *Then the following are equivalent:*

(1) f *is an epimorphism.*
(2) f *is the ring of definable scalars for* Φ *and* f *is left minimal.*
(3) *Restriction via* f *identifies* $\operatorname{Mod} S$ *with* $(\operatorname{Mod} R)_\Phi$.

PROOF. (1) $\Rightarrow$ (2) Given a map ϕ in $\operatorname{mod} R$, it has been shown in Lemma 11.15 that $\phi \otimes_R S$ is a split mono if and only if $\operatorname{Ker} T_\phi \in \mathcal{S}$ where $\mathcal{S}$ denotes the kernel of the exact functor $C(R) \to C(S)$ which extends f. Using this observation and the construction of the ring of definable scalars in Theorem 11.1, the assertion follows directly from the preceding proposition.

(2) $\Rightarrow$ (3) Suppose that $S = R_\Phi$. Using Theorem 11.16 and the assumption that $\phi \otimes_R S$ is a split mono for all $\phi \in \Phi$, we find a homomorphism $g\colon S \to S$ such that $f = g \circ f$ and the restriction via g of every S-module is Φ°-injective. But f is left minimal so that g is an isomorphism. Thus $\operatorname{Mod} S = (\operatorname{Mod} S)_{\Phi^\circ}$ is identified via restriction with the full subcategory of Φ-injective R-modules.

(3) $\Rightarrow$ (1) Restriction $\operatorname{Mod} S \to \operatorname{Mod} R$ being full implies that f is an epimorphism. □

REMARK 11.21. Identifying epimorphisms $R \to S$ and $R \to S'$ which coincide up to an isomorphism $S \to S'$, the epimorphisms starting in R form a set of cardinality at most $2^{\aleph_0 + \operatorname{card} R}$. This well-known fact is an immediate consequence of the preceding theorem.

We are now in a position to show that every epimorphism of rings is a localization.

COROLLARY 11.22. *A homomorphism* $f\colon R \to S$ *is an epimorphism if and only if there is a collection* Φ *of maps in* $\operatorname{mod} R$ *satisfying:*

(1) $\phi \otimes_R S$ *is a split mono for all* $\phi \in \Phi$.
(2) *If* $f'\colon R \to S'$ *is a homomorphism such that* $\phi \otimes_R S'$ *is a split mono for all* $\phi \in \Phi$, *then there is a unique homomorphism* $g\colon S \to S'$ *such that* $f' = g \circ f$.

PROOF. Combine Theorem 11.20 with Theorem 11.16. □

We suppose now for the rest of this section that $f\colon R \to S$ is an epimorphism. We use the notation $f^* = - \otimes_R S$ and denote by Φ the collection of maps ϕ in $\operatorname{mod} R$ such that $f^*(\phi)$ is a split mono in $\operatorname{mod} S$.

COROLLARY 11.23. *The ring $R[\Phi^-]$ of left fractions with respect to Φ is isomorphic to S via the map*

$$R[\Phi^-] \longrightarrow \operatorname{End}_S(S) = S, \quad [\alpha, \phi] \mapsto f^*(\phi)^- \circ f^*(\alpha)$$

where $f^*(\phi)^-$ denotes any left inverse of $f^*(\phi)$.

PROOF. The assertion is an immediate consequence of Theorem 11.20 and the description of R_Φ in Theorem 11.6. □

An analysis of the calculus of left fractions presented in Theorem 11.6 shows that any map $f^*(X) \to f^*(Y)$ in $\operatorname{mod} S$ can be expressed as a left fraction.

PROPOSITION 11.24. *For every map $\beta\colon f^*(X) \to f^*(Y)$ in $\operatorname{mod} S$ there are maps $\alpha\colon X \to Z$ and $\phi\colon Y \to Z$ in $\operatorname{mod} R$ such that $f^*(\phi)$ is a split mono and $\beta = f^*(\phi)^- \circ f^*(\alpha)$ for every left inverse $f^*(\phi)^-$ of $f^*(\phi)$.*

PROOF. Adapt the proof of Theorem 11.6. □

Given a matrix $A = [a_{ij}]$ with entries in R we define $f(A) = [f(a_{ij})]$.

COROLLARY 11.25. *For every $n \times m$ matrix B over S there are matrices A and F over R of appropriate size such that $B = f(F)^- \circ f(A)$ for some left inverse $f(F)^-$ of $f(F)$.*

PROOF. We view B as a map $\beta\colon f^*(R^n) \to f^*(R^m)$ and can apply the preceding proposition. Thus there are maps $\alpha\colon R^n \to X$ and $\phi\colon R^m \to X$ such that $\beta = f^*(\phi)^- \circ f^*(\alpha)$. Choosing an epi $\pi\colon R^l \to X$ we obtain maps $\alpha'\colon R^n \to R^l$ and $\phi'\colon R^m \to R^l$ such that $\alpha = \alpha' \circ \pi$ and $\phi = \phi' \circ \pi$. The matrices A and F corresponding to α' and ϕ', respectively, have the appropriate property. □

We are now in a position to show that the ring S is obtained from R by adjoining left inverses for elements of the form $\mathbf{x} = (x_1, \dots, x_n) \in R^n$. Given $\mathbf{x}, \mathbf{y} \in R^n$ we use the notation $\mathbf{xy} = \sum_{i=1}^n x_i y_i$ and $f(\mathbf{x}) = (f(x_1), \dots, f(x_n))$.

COROLLARY 11.26. *Let $s \in S$. Then there exist $\mathbf{r}, \mathbf{x} \in R^n$ and $f(\mathbf{x})^- \in S^n$ for some $n \in \mathbb{N}$ such that $1 = f(\mathbf{x})^- f(\mathbf{x})$ and $s = f(\mathbf{x})^- f(\mathbf{r})$.*

PROOF. Apply Corollary 11.25. □

COROLLARY 11.27. *Every finitely presented S-module is a direct summand of $X \otimes_R S$ for some finitely presented R-module X.*

PROOF. Every finitely presented S-module arises as cokernel of a map $\beta\colon S^n \to S^m$ which we can view as an $n \times m$ matrix over S. The assertion is now a consequence of Corollary 11.25. □

CHAPTER 12

Reflective definable subcategories

This chapter is devoted to definable subcategories of $\operatorname{Mod} R$ having the additional property that the inclusion functor has a left adjoint. We study also the rings of definable scalars which arise from such definable subcategories.

12.1. Reflective definable subcategories

We have seen in Theorem 11.20 that every epimorphism $R \to S$ identifies $\operatorname{Mod} S$ with a definable subcategory of $\operatorname{Mod} R$ which is reflective since the tensor functor $- \otimes_R S$ is a left adjoint of the restriction functor $\operatorname{Mod} S \to \operatorname{Mod} R$. In this section we study reflective definable subcategories of $\operatorname{Mod} R$ in more detail. Recall that a full subcategory of any category is *reflective* provided that the inclusion functor has a left adjoint. It is convenient to work within the framework of locally finitely presented categories in the sense of [**29**]. We recall briefly the relevant definitions.

Let $\mathcal{A}$ be an additive category which is cocomplete, i.e. $\mathcal{A}$ has arbitrary coproducts and cokernels. The category $\mathcal{A}$ is called *locally finitely presented* provided that the full subcategory $\operatorname{fp} \mathcal{A}$ of finitely presented objects is skeletally small and every object in $\mathcal{A}$ is a direct limit of finitely presented objects. Note that a locally finitely presented category is also complete and that $\operatorname{fp} \mathcal{A}$ is closed under cokernels.

Any locally finitely presented category $\mathcal{A}$ is equivalent to the category of left exact functors $(\operatorname{fp} \mathcal{A})^{\mathrm{op}} \to \mathrm{Ab}$ via the functor

$$\mathcal{A} \longrightarrow \operatorname{Lex}((\operatorname{fp} \mathcal{A})^{\mathrm{op}}, \mathrm{Ab}), \quad X \mapsto \operatorname{Hom}(-, X)|_{\operatorname{fp} \mathcal{A}}.$$

Conversely, for any skeletally small additive category $\mathcal{C}$ with cokernels the category $\mathcal{A} = \operatorname{Lex}(\mathcal{C}^{\mathrm{op}}, \mathrm{Ab})$ is locally finitely presented, and the Yoneda functor $X \mapsto \operatorname{Hom}(-, X)$ induces an equivalence $\mathcal{C} \to \operatorname{fp} \mathcal{A}$, see [**29**].

Given locally finitely presented categories $\mathcal{A}$ and $\mathcal{B}$ and a right exact functor $f \colon \operatorname{fp} \mathcal{A} \to \operatorname{fp} \mathcal{B}$, the restriction functor $\operatorname{Lex}((\operatorname{fp} \mathcal{B})^{\mathrm{op}}, \mathrm{Ab}) \to \operatorname{Lex}((\operatorname{fp} \mathcal{A})^{\mathrm{op}}, \mathrm{Ab})$, $X \mapsto X \circ f$, induces a functor $f_* \colon \mathcal{B} \to \mathcal{A}$; it has a left adjoint $f^* \colon \mathcal{A} \to \mathcal{B}$ which extends f. For example, the Yoneda functor

$$h \colon \operatorname{fp} \mathcal{A} \longrightarrow C(\mathcal{A}) = \operatorname{fp}(\operatorname{fp} \mathcal{A}, \mathrm{Ab})^{\mathrm{op}}, \quad X \mapsto \operatorname{Hom}(X, -)$$

extends to a fully faithful functor

$$d_{\mathcal{A}} \colon \mathcal{A} \simeq \operatorname{Lex}((\operatorname{fp} \mathcal{A})^{\mathrm{op}}, \mathrm{Ab}) \xrightarrow{h^*} \operatorname{Lex}(C(\mathcal{A})^{\mathrm{op}}, \mathrm{Ab}) = D(\mathcal{A})$$

which identifies $\mathcal{A}$ with the full subcategory $\operatorname{Ex}(C(\mathcal{A})^{\mathrm{op}}, \mathrm{Ab})$ of exact functors. This is a general version of Lemma 1.4 where $\mathcal{A} = \operatorname{Mod} R$.

The reflective subcategories which are closed under taking direct limits can be characterized as follows.

PROPOSITION 12.1. *Let $\mathcal{A}$ be a locally finitely presented category and suppose that $\mathcal{X}$ is a full subcategory of $\mathcal{A}$ which is closed under taking direct limits. Then the following are equivalent:*

(1) *The inclusion $\mathcal{X} \to \mathcal{A}$ has a left adjoint.*
(2) *$\mathcal{X}$ is locally finitely presented and there is a right exact functor $f\colon \operatorname{fp}\mathcal{A} \to \operatorname{fp}\mathcal{X}$ such that $f_*\colon \mathcal{X} \to \mathcal{A}$ is isomorphic to the inclusion.*

PROOF. (1) $\Rightarrow$ (2) The left adjoint, say $l\colon \mathcal{A} \to \mathcal{X}$, induces a functor $f\colon \operatorname{fp}\mathcal{A} \to \operatorname{fp}\mathcal{X}$ by Lemma 1.1; it is right exact since a left adjoint preserves cokernels. The category $\mathcal{X}$ is cocomplete since a left adjoint preserves colimits. If an object X in $\mathcal{X}$ is written as a direct limit $\varinjlim X_i$ of objects in $\operatorname{fp}\mathcal{A}$, then $X = l(X) \simeq \varinjlim f(X_i)$ since l preserves direct limits. Therefore $\mathcal{X}$ is locally finitely presented and we obtain $f^* = l$ since both functors coincide on $\operatorname{fp}\mathcal{A}$.

(2) $\Rightarrow$ (1) f^* is the left adjoint for f_*. □

REMARK 12.2. A reflective subcategory of $\mathcal{A}$ is complete since $\mathcal{A}$ is complete.

We are now in a position to characterize the definable subcategories which are reflective.

THEOREM 12.3. *Let $\mathcal{X}$ be a full subcategory of $\operatorname{Mod} R$ which is closed under taking direct limits. Then the following are equivalent:*

(1) *$\mathcal{X}$ is a definable subcategory which is reflective.*
(2) *$\mathcal{X}$ is reflective.*
(3) *$\mathcal{X}$ is locally finitely presented and closed under taking kernels and products.*
(4) *$\mathcal{X}$ is locally finitely presented. Moreover, there is a ring homomorphism $R \to \operatorname{End}_R(X)$ for some $X \in \operatorname{fp}\mathcal{X}$ and a functorial isomorphism $M \simeq \operatorname{Hom}_R(X, M)$ for all M in $\mathcal{X}$.*

PROOF. (1) $\Rightarrow$ (2) Trivial.

(2) $\Rightarrow$ (3) $\mathcal{X}$ is locally finitely presented by the preceding proposition. It is closed under kernels and products since a right adjoint preserves limits.

(3) $\Rightarrow$ (1) Using Freyd's adjoint functor theorem one shows that the inclusion has a left adjoint $l\colon \operatorname{Mod} R \to \mathcal{X}$, see [**29**]. The left adjoint restricts to a right exact functor $\operatorname{mod} R \to \operatorname{fp}\mathcal{X}$ by Lemma 1.1 which extends to an exact functor $f\colon C(R) \simeq \operatorname{fp}(\operatorname{mod} R, \mathrm{Ab})^{\mathrm{op}} \to C(\mathcal{X})$. We claim that $\operatorname{Ker} f = \mathcal{S}_{\mathcal{X}}$, i.e. $\mathcal{X}$ is the full subcategory of R-modules M such that $\operatorname{Hom}(\operatorname{Ker} f, T_M) = 0$. Identifying $\mathcal{X}$ with the image of $d_{\mathcal{X}}\colon \mathcal{X} \to D(\mathcal{X})$ and $\operatorname{Mod} R$ with the image of $\operatorname{Mod} R \to D(R)$, $M \mapsto T_M$, the functor f^* extends l and the right adjoint f_* extends the inclusion $\mathcal{X} \to \operatorname{Mod} R$. Applying Lemma 11.18, f^* induces an equivalence $D(R)/\operatorname{Ker} f^* \to D(\mathcal{X})$ and therefore the inclusion $\mathcal{X} \to \operatorname{Mod} R$ identifies $\mathcal{X}$ with the full subcategory of R-modules M satisfying $\operatorname{Hom}(\operatorname{Ker} f^*, T_M) = 0$. The last condition is equivalent to $\operatorname{Hom}(\operatorname{Ker} f, T_M) = 0$ since $\operatorname{Ker} f^* = \varinjlim \operatorname{Ker} f$, and therefore $\mathcal{X}$ is definable by Theorem 2.1.

(2) $\Leftrightarrow$ (4) We need to show that the condition (4) is equivalent to condition (2) in the preceding proposition. In fact, a right exact functor $f\colon \operatorname{mod} R \to \operatorname{fp}\mathcal{X}$ with $X = f(R)$ is completely determined by the homomorphism $R \to \operatorname{End}_R(X)$ induced by f, and $f_*(M) = \operatorname{Hom}_R(X, M)$ for all M in $\mathcal{X}$. □

We now describe the collection Φ of maps corresponding to a reflective definable subcategory and we show that the homomorphism $R \to \operatorname{End}_R(X)$ occuring in condition (4) is precisely the ring of definable scalars for Φ.

COROLLARY 12.4. *Let $\mathcal{X}$ be a reflective definable subcategory of* $\operatorname{Mod} R$. *Denote by $l\colon \operatorname{Mod} R \to \mathcal{X}$ a left adjoint of the inclusion and let Φ be the collection of maps ϕ in* $\operatorname{mod} R$ *such that $l(\phi)$ is a split mono. Then $\mathcal{X} = (\operatorname{Mod} R)_\Phi$, and the homomorphism $R \to \operatorname{End}_R(l(R))$ induced by l is the ring of definable scalars for Φ.*

PROOF. In part (3) $\Rightarrow$ (1) of the preceding proof $\mathcal{S}_\mathcal{X}$ is computed. It is easily checked that $\operatorname{Ker} T_\phi \in \mathcal{S}_\mathcal{X}$ if and only if $l(\phi)$ is a split mono. Therefore $\mathcal{X} = (\operatorname{Mod} R)_\Phi$. Identifying $(\operatorname{Mod} R)_\Phi$ with the full subcategory of fp-injective objects in $D(R)/\mathcal{T}_\Phi$ as in Corollary 2.4, we have $q_\Phi(T_M) \simeq l(M)$ for every R-module M since the quotient functor $q_\Phi : D(R) \to D(R)/\mathcal{T}_\Phi$ is a left adjoint of the section functor s_Φ which induces the inclusion $(\operatorname{Mod} R)_\Phi \to \operatorname{Mod} R$. By definition, $R_\Phi = \operatorname{End}(q_\Phi(T_R))$ and the assertion follows. ☐

EXAMPLE 12.5. Let $(\mathcal{T}, \mathcal{F})$ be a torsion theory for $\operatorname{Mod} R$. Then the torsion-free modules form a reflective definable subcategory if and only if $\mathcal{F}$ is closed under direct limits.

12.2. Φ-continuous modules

In this section we study a special class of definable subcategories of a module category $\operatorname{Mod} R$. We begin with some definitions. Let $\phi\colon X \to Y$ be a map of R-modules. An R-module M is called *ϕ-continuous* if the induced map $\operatorname{Hom}_R(Y, M) \to \operatorname{Hom}_R(X, M)$ is bijective. If Φ is a collection of maps, then M is Φ-continuous if M is ϕ-continuous for every ϕ in Φ. Given a map $\phi\colon X \to Y$ of R-modules, we define $\widetilde{\phi}$ to be the map $[\phi\, 0]\colon X \coprod \operatorname{Coker}\phi \to Y$ and we write $\widetilde{\Phi} = \{\widetilde{\phi} \mid \phi \in \Phi\}$ for every collection Φ of maps. It is clear that any R-module is ϕ-continuous if and only if it is $\widetilde{\phi}$-injective.

The concept of a Φ-continuous object makes sense for any category $\mathcal{A}$ and for any collection of maps Φ in $\mathcal{A}$. In fact, it is convenient to follow Gabriel and Ulmer [**29**]. We shall discuss the properties of the full subcategory $\mathcal{A}_{\widetilde{\Phi}}$ formed by the Φ-continuous objects under the assumption that $\mathcal{A}$ is a locally finitely presented additive category. We begin with some calculus of fractions which is based on [**30**].

PROPOSITION 12.6. *Let $\mathcal{C}$ be an additive category with cokernels and let Φ be a collection of maps in $\mathcal{C}$. Then there exists an additive category $\mathcal{C}[\Phi^{-1}]$ with cokernels and a right exact functor $p\colon \mathcal{C} \to \mathcal{C}[\Phi^{-1}]$ such that $p(\phi)$ is invertible for all $\phi \in \Phi$. Furthermore:*

(1) *$\mathcal{C}$ and $\mathcal{C}[\Phi^{-1}]$ have the same objects and p acts on them as identity.*
(2) *If $f\colon \mathcal{C} \to \mathcal{D}$ is a right exact functor into an additive category with cokernels such that $f(\phi)$ is invertible for all $\phi \in \Phi$, then there is a unique right exact functor $g\colon \mathcal{C}[\Phi^{-1}] \to \mathcal{D}$ such that $f = g \circ p$.*

The proof of this proposition depends on the following concept. We say that a collection Φ of maps in $\mathcal{C}$ *admits a calculus of left fractions* if

(C1) $\mathrm{id}_X \in \Phi$ for all objects X in $\mathcal{C}$;
(C2) if $\phi\colon X \to Y$ and $\psi\colon Y \to Z$ are in Φ, then $\psi \circ \phi \in \Phi$;
(C3) if $\phi\colon X \to Y$ is in Φ and

$$\begin{array}{ccc} X & \longrightarrow & Z \\ \downarrow{\scriptstyle\phi} & & \downarrow{\scriptstyle\psi} \\ Y & \longrightarrow & P \end{array}$$

is a pushout diagram, then $\psi \in \Phi$;

(C4) if $\phi \in \Phi$, then $\operatorname{Coker}\phi \to 0$ is in Φ.

Note that the conditions (C1) – (C4) imply the conditions given in [**30**].

Proof of Proposition 12.6. Denote by $\overline{\Phi}$ the smallest collection of maps in $\mathcal{C}$ which admits a calculus of left fractions and contains Φ. Let $\mathcal{C}[\overline{\Phi}^{-1}]$ be the category of fractions which is obtained from $\mathcal{C}$ by formally adjoining inverses for each map in $\overline{\Phi}$, see [**30**] and define, by abuse of notation, $\mathcal{C}[\Phi^{-1}] = \mathcal{C}[\overline{\Phi}^{-1}]$. The category $\mathcal{C}[\Phi^{-1}]$ is additive, has cokernels and there is a right exact functor $p\colon \mathcal{C} \to \mathcal{C}[\Phi^{-1}]$ having the appropriate universal property, i.e. every functor $f\colon \mathcal{C} \to \mathcal{D}$ inverting $\overline{\Phi}$ induces a unique functor $g\colon \mathcal{C}[\Phi^{-1}] \to \mathcal{D}$ such that $f = g \circ p$, see [**30**]; in particular (1) holds. In order to verify (2) observe that for any right exact functor $f\colon \mathcal{C} \to \mathcal{D}$ the collection Ψ of maps in $\mathcal{C}$ which become invertible in $\mathcal{D}$ admits a calculus of left fractions. Therefore $\overline{\Phi} \subseteq \Psi$ if $\Phi \subseteq \Psi$, and we conclude that the functor p has the universal property which is stated in (2). □

Theorem 12.7. *Let $\mathcal{A}$ be a locally finitely presented category and let Φ be a collection of maps in $\operatorname{fp}\mathcal{A}$. Then $\mathcal{A}_{\tilde{\Phi}}$ is a locally finitely presented category and the inclusion $\mathcal{A}_{\tilde{\Phi}} \to \mathcal{A}$ has a left adjoint which induces an equivalence between $(\operatorname{fp}\mathcal{A})[\Phi^{-1}]$ and $\operatorname{fp}\mathcal{A}_{\tilde{\Phi}}$.*

Proof. We use the basic facts about locally finitely presented categories from the preceding section. Consider $\mathcal{C} = \operatorname{fp}\mathcal{A}$ and the functor $p\colon \mathcal{C} \to \mathcal{C}[\Phi^{-1}]$ which induces a fully faithful functor $p_*\colon \operatorname{Lex}(\mathcal{C}[\Phi^{-1}]^{\mathrm{op}}, \mathrm{Ab}) \to \operatorname{Lex}(\mathcal{C}^{\mathrm{op}}, \mathrm{Ab})$, $X \mapsto X \circ p$. If we identify $\mathcal{A} = \operatorname{Lex}(\mathcal{C}^{\mathrm{op}}, \mathrm{Ab})$, then p_* induces an equivalence between the locally finitely presented category $\operatorname{Lex}(\mathcal{C}[\Phi^{-1}]^{\mathrm{op}}, \mathrm{Ab})$ and $\mathcal{A}_{\tilde{\Phi}}$. Moreover, there is a unique functor $p^*\colon \operatorname{Lex}(\mathcal{C}^{\mathrm{op}}, \mathrm{Ab}) \to \operatorname{Lex}(\mathcal{C}[\Phi^{-1}]^{\mathrm{op}}, \mathrm{Ab})$ which preserves direct limits and extends the functor $\operatorname{Hom}(-, X) \mapsto \operatorname{Hom}(-, p(x))$ which is defined for every X in $\mathcal{C}$. It is not hard to see that p^* is a left adjoint for p_* and therefore the proof is complete. □

We now collect the basic properties of the category of Φ-continuous modules.

Corollary 12.8. *Let Φ be a collection of maps in $\operatorname{mod} R$. Then the definable subcategory $\mathcal{X} = (\operatorname{Mod} R)_{\tilde{\Phi}}$ formed by the Φ-continuous R-modules has the following properties:*

1. *$\mathcal{X}$ has kernels, cokernels, products and coproducts.*
2. *Every object in $\mathcal{X}$ is a direct limit of finitely presented objects.*
3. *The inclusion $\mathcal{X} \to \operatorname{Mod} R$ has a left adjoint and preserves kernels, products and direct limits.*
4. *The left adjoint of the inclusion functor restricts to a right exact functor $p\colon \operatorname{mod} R \to \operatorname{fp}\mathcal{X}$.*
5. *p induces an equivalence $(\operatorname{mod} R)[\Phi^{-1}] \to \operatorname{fp}\mathcal{X}$.*

Proof. We can apply the preceding theorem since any module category is locally finitely presented. In fact, the assertions hold in any locally finitely presented category. □

The ring of definable scalars $R_{\tilde{\Phi}}$ can be computed as follows. Suppose that Φ admits a calculus of left fractions (if not, replace Φ by the smallest collection of maps which contains Φ and admits a calculus of left fractions) and denote by

$R(\Phi^{-1})$ the collection of pairs (α, ϕ) of maps $\alpha, \phi \colon R \to X$ in $\operatorname{mod} R$ such that $\phi \in \Phi$. Define an equivalence relation on $R(\Phi^{-1})$ as in Theorem 11.6 and denote by $R[\Phi^{-1}]$ the set of equivalence classes, endowed with addition and multiplication as in Theorem 11.6. We denote by $l \colon \operatorname{Mod} R \to (\operatorname{Mod} R)_{\tilde{\Phi}}$ the left adjoint of the inclusion $(\operatorname{Mod} R)_{\tilde{\Phi}} \to \operatorname{Mod} R$.

PROPOSITION 12.9. *The assignment* $[\alpha, \phi] \mapsto l(\phi)^{-1} \circ l(\alpha)$ *induces an isomorphism* $R[\Phi^{-1}] \to R_{\tilde{\Phi}}$.

PROOF. The map $[\alpha, \phi] \mapsto l(\phi)^{-1} \circ l(\alpha)$ induces an iso between $R[\Phi^{-1}]$ and the endomorphism ring of $l(R)$, see [**30**]. Identifying $\operatorname{End}_R(l(R))$ with $R_{\tilde{\Phi}}$ as in Corollary 12.4 the assertion follows. □

We end this section with some examples of Φ-continuous modules.

EXAMPLE 12.10. Let $\mathcal{S}$ be a class of R-modules and S be a subset of R.

(1) Let $\Phi = \{X \to 0 \mid X \in \mathcal{S}\}$. Then an R-module M is Φ-continuous if and only if M is Φ-injective if and only if $\operatorname{Hom}_R(\mathcal{S}, M) = 0$.

(2) Choose for each X in $\mathcal{S}$ an exact sequence $0 \to \Omega_X \overset{\phi_X}{\to} P_X \to X \to 0$ with P_X projective and let $\Phi = \{\phi_X \mid X \in \mathcal{S}\}$. Then the perpendicular category $\mathcal{S}^{\perp}$ of R-modules M satisfying $\operatorname{Hom}_R(\mathcal{S}, M) = 0 = \operatorname{Ext}^1_R(\mathcal{S}, M)$ is precisely $(\operatorname{Mod} R)_{\tilde{\Phi}}$. If each module in $\mathcal{S}$ is finitely presented and R is right coherent, then there is a choice for Φ such that each map in Φ belongs to $\operatorname{mod} R$.

(3) Let $\Phi = \{R \overset{s}{\to} R \mid s \in S\}$. Then an R-module M is Φ-continuous if and only if for every $m \in M$ and $s \in S$ there is a unique $n \in M$ such that $m = n \cdot s$. The usual ring of fractions $R[S^{-1}]$ is precisely $R[\Phi^{-1}]$, and the restriction functor $\operatorname{Mod} R[\Phi^{-1}] \to \operatorname{Mod} R$ identifies $\operatorname{Mod} R[\Phi^{-1}]$ with $(\operatorname{Mod} R)_{\tilde{\Phi}}$.

12.3. Definable and universal localizations

Let Φ be a collection of maps in $\operatorname{mod} R$. We call a ring homomorphism $f \colon R \to S$ the *definable* (respectively, *universal*) *localization* with respect to Φ if the following conditions are satisfied:

(L1) $\phi \otimes_R S$ is a split mono (respectively, iso) in $\operatorname{Mod} S$ for all $\phi \in \Phi$.

(L2) If $f' \colon R \to S'$ is a ring homomorphism such that $\phi \otimes_R S'$ is a split mono (respectively, iso) in $\operatorname{Mod} S'$ for all $\phi \in \Phi$, then there is a unique homomorphism $g \colon S \to S'$ such that $f' = g \circ f$.

It is clear that such a localization is unique up to an isomorphism if it exists. Moreover, it is an epimorphism of rings. Given a definable (respectively, universal) localization $f \colon R \to S$ with respect to Φ, it follows immediately that the restriction of every S-module via f becomes a Φ-injective (respectively, Φ-continuous) R-module. We call the localization *strong* if in addition the following condition holds:

(L3) If M is Φ-injective (respectively, Φ-continuous), then M is the restriction of an S-module.

Our definition is motivated by Schofield's universal localization [**75**] and by Corollary 11.22 which says that every epimorphism of rings is a strong definable localization. We shall now discuss the existence of both localizations for fixed Φ.

THEOREM 12.11. *For a collection* Φ *of maps in* $\operatorname{mod} R$ *the following conditions are equivalent.*

(1) *The strong definable localization with respect to* Φ *exists.*

(2) *$f_\Phi\colon R \to R_\Phi$ is the strong definable localization with respect to Φ.*
(3) *Restriction via f_Φ induces an equivalence* $\operatorname{Mod} R_\Phi \to (\operatorname{Mod} R)_\Phi$.
(4) *The inclusion* $(\operatorname{Mod} R)_\Phi \to \operatorname{Mod} R$ *is right exact and has a left adjoint.*
(5) *$\phi \otimes_R R_\Phi$ is a split mono for all $\phi \in \Phi$ and f_Φ is left minimal.*

PROOF. (1) $\Rightarrow$ (2) The strong definable localization with respect to Φ, say $R \to S$, is an epimorphism which identifies via restriction $\operatorname{Mod} S$ with the category of Φ-injective R-modules because of condition (L3). Therefore the exact functor $C(R) \to C(S)$ factors through the quotient functor $C(R) \to C(R)/\mathcal{S}_\Phi$ and induces an equivalence $p\colon C(R)/\mathcal{S}_\Phi \to C(S)$ by Lemma 11.2. The functor p induces an isomorphism $R_\Phi \to S$.

(2) $\Rightarrow$ (3) Restriction via f_Φ induces a functor $\operatorname{Mod} R_\Phi \to (\operatorname{Mod} R)_\Phi$ which is fully faithful since f_Φ is an epimorphism. It is dense by condition (L3).

(3) $\Rightarrow$ (4) Restriction via f_Φ is right exact and has a left adjoint. Condition (3) therefore implies (4).

(4) $\Rightarrow$ (5) Suppose there is a left adjoint $l\colon \operatorname{Mod} R \to \mathcal{X} = (\operatorname{Mod} R)_\Phi$ and let $S = \operatorname{End}_R(l(R))$. We claim that the homomorphism $f\colon R \to S$ induced by l is the ring of definable scalars for Φ. Observe first that $\mathcal{X}$ is an abelian category with an exact inclusion. More precisely, we use that $\mathcal{X}$ is a full subcategory with kernels and cokernels, and that the inclusion is left exact and right exact. This observation implies that $l(R)$ is a projective object in $\mathcal{X}$ since R is projective in $\operatorname{Mod} R$ and the right adjoint of l is exact. Furthermore, $l(R)$ is finitely presented in $\mathcal{X}$ by Lemma 1.1 since R is finitely presented in $\operatorname{Mod} R$ and the right ajoint preserves direct limits. Finally, $l(R)$ is a generator for $\mathcal{X}$ and therefore $\mathcal{X}$ is equivalent to $\operatorname{Mod} S$ via $M \mapsto \operatorname{Hom}_R(l(R), M)$. It follows that restriction via f induces an equivalence $\operatorname{Mod} S \to \mathcal{X}$ and the argument used in the proof of (1) $\Rightarrow$ (2) shows that f is the ring of definable scalars for Φ. Also, f is an epimorphism since the restriction functor is full, and therefore f is left minimal. Finally, $\phi \otimes_R S$ is a split mono for all ϕ by Lemma 11.15.

(5) $\Rightarrow$ (1) We show that f_Φ is the strong definable localization with respect to Φ. (L2) follows from Theorem 11.16 and (L3) follows from the implication (2) $\Rightarrow$ (3) in Theorem 11.20 (or its proof). □

An easy observation shows that definable and universal localizations are closely related.

LEMMA 12.12. *Let Φ be collection of maps in* $\operatorname{mod} R$. *A ring homomorphism $f\colon R \to S$ is the (strong) universal localization with respect to Φ if and only if f is the (strong) definable localization with respect to $\widetilde{\Phi}$.*

PROOF. Clear. □

COROLLARY 12.13. *For a collection Φ of maps in* $\operatorname{mod} R$ *the following conditions are equivalent.*

(1) *The strong universal localization with respect to Φ exists.*
(2) *$f_{\widetilde{\Phi}}\colon R \to R_{\widetilde{\Phi}}$ is the strong universal localization with respect to Φ.*
(3) *Restriction via $f_{\widetilde{\Phi}}$ induces an equivalence* $\operatorname{Mod} R_{\widetilde{\Phi}} \to (\operatorname{Mod} R)_{\widetilde{\Phi}}$.
(4) *The inclusion* $(\operatorname{Mod} R)_{\widetilde{\Phi}} \to \operatorname{Mod} R$ *is right exact.*
(5) *$\phi \otimes_R R_{\widetilde{\Phi}}$ is an iso for all $\phi \in \Phi$ and $f_{\widetilde{\Phi}}$ is left minimal.*

COROLLARY 12.14. *Let $R \to R_{\widetilde{\Phi}}$ be the strong universal localization with respect to Φ. Then the functor $- \otimes_R R_{\widetilde{\Phi}}\colon$* $\operatorname{mod} R \to \operatorname{mod} R_{\widetilde{\Phi}}$ *induces an equivalence*

$(\operatorname{mod} R)[\Phi^{-1}] \to \operatorname{mod} R_{\tilde{\Phi}}$. *In particular* $R[\Phi^{-1}] \simeq R_{\tilde{\Phi}}$, *and every finitely presented* $R_{\tilde{\Phi}}$*-module is of the form* $X \otimes R_{\tilde{\Phi}}$ *for some finitely presented* R*-module* X.

PROOF. Use Corollary 12.8 and Proposition 12.9. □

In [**75**], Schofield constructs the universal localization with respect to a collection of maps between finitely generated projective modules. We recover Schofield's localization and some of its properties discussed in [**75**] from the preceding corollaries.

COROLLARY 12.15. *Let* Φ *be a collection of maps between finitely generated projective* R*-modules. Then* $f_{\tilde{\Phi}} \colon R \to R_{\tilde{\Phi}}$ *is the strong universal localization with respect to* Φ.

PROOF. It is clear that $(\operatorname{Mod} R)_{\tilde{\Phi}}$ is a subcategory of $\operatorname{Mod} R$ which is closed under taking cokernels. The assertion is therefore a consequence of Corollary 12.13. □

With some extra assumptions, there is a convenient description of $\operatorname{Mod} R_{\tilde{\Phi}}$ using perpendicular categories. We present a result which is more or less explicitly contained in [**76, 33, 16**].

Suppose there is given a collection Φ of maps in $\operatorname{proj} R$ such that the projective dimension of $\operatorname{Coker} \phi$ is bounded by 1 for all $\phi \in \Phi$. Then it is easily checked that an R-module M is Φ-continuous if and only if M belongs to the perpendicular category $\mathcal{S}^{\perp}$ of R-modules N satisfying $\operatorname{Hom}_R(\mathcal{S}, N) = 0 = \operatorname{Ext}^1_R(\mathcal{S}, N)$ where $\mathcal{S} = \{\operatorname{Ker} \phi, \operatorname{Coker} \phi \mid \phi \in \Phi\}$. This observation in combination with the preceding result has the following consequence.

THEOREM 12.16. *For a full subcategory* $\mathcal{X}$ *of* $\operatorname{Mod} R$ *the following are equivalent:*

(1) $\mathcal{X} = \mathcal{S}^{\perp}$ *for some collection* $\mathcal{S}$ *of objects in* $\operatorname{mod} R$ *with* $\operatorname{pd} \mathcal{S} \leq 1$.
(2) $\mathcal{X} = (\operatorname{Mod} R)_{\tilde{\Phi}} \simeq \operatorname{Mod} R_{\tilde{\Phi}}$ *for some collection* Φ *of maps in* $\operatorname{proj} R$ *with* $\operatorname{pd}\{\operatorname{Coker} \phi \mid \phi \in \Phi\} \leq 1$.

PROOF. For any class $\mathcal{C}$ of R-modules we denote by $\operatorname{pd} \mathcal{C}$ the supremum of the projective dimensions of the modules in $\mathcal{C}$. The proof uses the fact that for any strong universal localization $R \to R_{\tilde{\Phi}}$ with respect to Φ the restriction functor identifies $\operatorname{Mod} R_{\tilde{\Phi}}$ with the full subcategory of Φ-continuous R-modules.

(1) $\Rightarrow$ (2) Choose for each X in $\mathcal{S}$ a mono ϕ_X lying in $\operatorname{proj} R$ with $\operatorname{Coker} \phi_X = X$ and let $\Phi = \{\phi_X \mid X \in \mathcal{S}\}$.

(2) $\Rightarrow$ (1) Choose $\mathcal{S} = \{\operatorname{Ker} \phi, \operatorname{Coker} \phi \mid \phi \in \Phi\}$. □

We give an application of the preceding theorem.

COROLLARY 12.17. *Let* R *be a right hereditary and noetherian ring and suppose that* $\mathcal{T}$ *is a localizing subcategory of* $\operatorname{Mod} R$. *Then there exists a universal localization* $R \to R_{\tilde{\Phi}}$ *with respect to some collection* Φ *of maps in* $\operatorname{proj} R$ *such that* $- \otimes_R R_{\tilde{\Phi}} \colon \operatorname{Mod} R \to \operatorname{Mod} R_{\tilde{\Phi}}$ *induces an equivalence between* $\operatorname{Mod} R/\mathcal{T}$ *and* $\operatorname{Mod} R_{\tilde{\Phi}}$.

PROOF. $\mathcal{T} = \varinjlim \mathcal{S}$ for some Serre subcategory $\mathcal{S}$ of $\operatorname{mod} R$ since R is right noetherian, and $\operatorname{pd} \mathcal{S} \leq 1$ since R is right hereditary. Thus the section functor $\operatorname{Mod} R/\mathcal{T} \to \operatorname{Mod} R$ induces an equivalence between $\operatorname{Mod} R/\mathcal{T}$ and $\mathcal{S}^{\perp}$ by Proposition A.6, and the assertion now follows from Theorem 12.16. □

CHAPTER 13

Sheaves

The aim in this chapter is to exhibit the geometric nature of the Ziegler spectrum. To this end we introduce sheafs of local morphisms for any locally coherent category. Then we apply this construction to the locally coherent category $D(R)$. Using the Zariski topology on $\operatorname{Ind} R$ we obtain from the ring of definable scalars a sheaf of rings $\mathcal{O}_R$ and assign to every R-module M a sheaf $\widetilde{M}$.

13.1. Sheaves on the Gabriel spectrum

Throughout this section we fix a locally coherent Grothendieck category $\mathcal{A}$. Recall that a Grothendieck category $\mathcal{A}$ is *locally coherent* if the full subcategory $\mathcal{C} = \operatorname{fp} \mathcal{A}$ of finitely presented objects is abelian and every object in $\mathcal{A}$ is a direct limit of finitely presented objects. Denote by $\mathbf{X} = \operatorname{Sp} \mathcal{A}$ the *Gabriel spectrum* of $\mathcal{A}$, i.e. the set of isomorphism classes of indecomposable injective objects in $\mathcal{A}$. Given an object X in $\mathcal{C}$, let $\mathbf{U}_X = \{P \in \mathbf{X} \mid \operatorname{Hom}(X, P) = 0\}$. The collection of subsets $(\mathbf{U}_X)_{X \in \mathcal{C}}$ forms a basis of open subsets for the (Zariski) topology on $\operatorname{Sp} \mathcal{A}$ which is closed under finite intersections since $\mathbf{U}_X \cap \mathbf{U}_Y = \mathbf{U}_{X \coprod Y}$.

In this section we present a construction which assigns to each pair of objects M, N in $\mathcal{A}$ a sheaf of abelian groups $\widetilde{\operatorname{Hom}}_{\mathbf{X}}(M, N)$ on $\mathbf{X}$. We shall use freely the localization theory for abelian categories and some additional facts about locally coherent categories which can be found in the appendix. We refer the reader to [**34**] for basic facts about sheaves. The category of (pre)sheaves of abelian groups on $\mathbf{X}$ will be denoted by $\operatorname{Presh} \mathbf{X}$ and $\operatorname{Sh} \mathbf{X}$, respectively.

We need some notation. Given an object X in $\mathcal{C}$, we denote by $\mathcal{S}_X$ the smallest Serre subcategory of $\mathcal{C}$ containing X, and $\mathcal{T}_X$ denotes the smallest localizing subcategory of $\mathcal{A}$ containing X. Note that $\mathcal{T}_X = \varinjlim \mathcal{S}_X$.

Let us begin with some preliminary lemmas. Suppose that $\mathcal{S}$ is a Serre subcategory of $\mathcal{C}$ and denote by $\mathcal{T} = \varinjlim \mathcal{S}$ the full subcategory of $\mathcal{A}$ formed by the direct limits of objects in $\mathcal{S}$. Note that $\mathcal{T}$ is a localizing subcategory of $\mathcal{A}$.

LEMMA 13.1. *Let $\mathcal{S} = \bigcup \mathcal{S}_i$ be the directed union of Serre subcategories of $\mathcal{C}$. If X and Y are objects in $\mathcal{C}$, then $\varinjlim \operatorname{Hom}_{\mathcal{C}/\mathcal{S}_i}(X, Y) \simeq \operatorname{Hom}_{\mathcal{C}/\mathcal{S}}(X, Y)$.*

PROOF. Left to the reader. □

LEMMA 13.2. *Let X be a finitely presented object in $\mathcal{A}$.*

(1) *X is a finitely presented object in $\mathcal{A}/\mathcal{T}$, i.e. $\operatorname{Hom}_{\mathcal{A}/\mathcal{T}}(X, -)$ commutes with direct limits.*
(2) *The inclusion $\mathcal{C} \to \mathcal{A}$ induces a fully faithful functor $\mathcal{C}/\mathcal{S} \to \mathcal{A}/\mathcal{T}$.*

PROOF. See Proposition A.5. □

LEMMA 13.3. *Let M and N be objects in $\mathcal{A}$ and suppose that M is finitely presented. Then*

$$\varinjlim_{X\in\mathcal{S}} \operatorname{Hom}_{\mathcal{A}/\mathcal{T}_X}(M,N) \simeq \operatorname{Hom}_{\mathcal{A}/\mathcal{T}}(M,N).$$

PROOF. Write $N = \varinjlim_{i\in I} N_i$ as a direct limit of finitely presented objects. Using the preceding lemmas we obtain

$$\begin{aligned}\varinjlim_{X\in\mathcal{S}} \operatorname{Hom}_{\mathcal{A}/\mathcal{T}_X}(M,N) &\simeq \varinjlim_{X\in\mathcal{S}} \varinjlim_{i\in I} \operatorname{Hom}_{\mathcal{A}/\mathcal{T}_X}(M,N_i)\\ &\simeq \varinjlim_{X\in\mathcal{S}} \varinjlim_{i\in I} \operatorname{Hom}_{\mathcal{C}/\mathcal{S}_X}(M,N_i)\\ &\simeq \varinjlim_{i\in I} \operatorname{Hom}_{\mathcal{C}/\mathcal{S}}(M,N_i)\\ &\simeq \operatorname{Hom}_{\mathcal{A}/\mathcal{T}}(M,N).\end{aligned}$$

□

Now fix a pair M, N of objects in $\mathcal{A}$. We define the *presheaf of local morphisms* $\operatorname{Hom}_{\mathbf{X}}(M,N)$ on $\mathbf{X}$ as follows. Let $\operatorname{Hom}_{\mathbf{X}}(M,N)(\mathbf{U}_X) = \operatorname{Hom}_{\mathcal{A}/\mathcal{T}_X}(M,N)$ for every basic open set $\mathbf{U}_X$. If $\mathbf{U}_Y \subseteq \mathbf{U}_X$, then $\mathcal{T}_X \subseteq \mathcal{T}_Y$ by Proposition A.9, and the quotient functor $\mathcal{A}/\mathcal{T}_X \to \mathcal{A}/\mathcal{T}_Y$ induces the restriction morphism

$$\rho_{\mathbf{U}_X\mathbf{U}_Y} \colon \operatorname{Hom}_{\mathbf{X}}(M,N)(\mathbf{U}_X) \longrightarrow \operatorname{Hom}_{\mathbf{X}}(M,N)(\mathbf{U}_Y).$$

If $\mathbf{U}$ is an arbitrary open set, then we define

$$\operatorname{Hom}_{\mathbf{X}}(M,N)(\mathbf{U}) = \varprojlim_{\mathbf{U}_X\subseteq\mathbf{U}} \operatorname{Hom}_{\mathbf{X}}(M,N)(\mathbf{U}_X).$$

For $P \in \mathbf{X}$, the stalk $\operatorname{Hom}_{\mathbf{X}}(M,N)_P$ is the direct limit $\varinjlim_{P\in\mathbf{U}_X} \operatorname{Hom}_{\mathbf{X}}(M,N)(\mathbf{U}_X)$ since the basis $(\mathbf{U}_X)_{X\in\mathcal{C}}$ is closed under finite intersections. The stalk can be described as follows. Let $\mathcal{A}_P = \mathcal{A}/\mathcal{T}$ where $\mathcal{T} = \varinjlim \mathcal{S}$ and $\mathcal{S}$ denotes the Serre subcategory of all objects X in $\mathcal{C}$ with $\operatorname{Hom}(X,P) = 0$.

LEMMA 13.4. *There is a natural morphism* $\operatorname{Hom}_{\mathbf{X}}(M,N)_P \to \operatorname{Hom}_{\mathcal{A}_P}(M,N)$ *which is an isomorphism if M is finitely presented.*

PROOF. The assertion follows from Lemma 13.3 since $X \in \mathcal{S}$ if and only if $P \in \mathbf{U}_X$. □

LEMMA 13.5. *The natural map* $\operatorname{Hom}_{\mathbf{X}}(M,N)(\mathbf{U}) \to \prod_{P\in\mathbf{U}} \operatorname{Hom}_{\mathbf{X}}(M,N)_P$ *is injective.*

PROOF. We may assume that $\mathbf{U} = \mathbf{U}_X$. The category $\mathcal{A}/\mathcal{T}_X$ is locally coherent and therefore $\mathbf{U}_X = \operatorname{Sp}\mathcal{A}/\mathcal{T}_X$ cogenerates $\mathcal{A}/\mathcal{T}_X$, see Proposition A.9. It follows that $\operatorname{Hom}_{\mathcal{A}/\mathcal{T}_X}(\phi,P) \neq 0$ for some $P \in \mathbf{U}_X$ if $0 \neq \phi \in \operatorname{Hom}_{\mathbf{X}}(M,N)(\mathbf{U}_X)$. The quotient functor $\mathcal{A}/\mathcal{T}_X \to \mathcal{A}_P$ induces an isomorphism $\operatorname{Hom}_{\mathcal{A}/\mathcal{T}_X}(L,P) \simeq \operatorname{Hom}_{\mathcal{A}_P}(L,P)$ for every object L. Thus the image of ϕ under the composition $\operatorname{Hom}_{\mathbf{X}}(M,N)(\mathbf{U}) \to \operatorname{Hom}_{\mathbf{X}}(M,N)_P \to \operatorname{Hom}_{\mathcal{A}_P}(M,N)$ is non-zero. □

We collect some basic properties of the presheaf $\operatorname{Hom}_{\mathbf{X}}(M,N)$. To this end recall that a presheaf F on $\mathbf{X}$ is *seperated* if the natural morphism $F(\mathbf{U}) \to \prod_{i\in I} F(\mathbf{U}_i)$ is injective for every family $(\mathbf{U}_i)_{i\in I}$ of open sets with $\mathbf{U} = \bigcup_{i\in I} \mathbf{U}_i$.

PROPOSITION 13.6. *The functor* $\operatorname{Hom}_{\mathbf{X}}(-,-) \colon \mathcal{A}^{\mathrm{op}} \times \mathcal{A} \longrightarrow \operatorname{Presh}\mathbf{X}$ *is left exact in both arguments and every presheaf* $\operatorname{Hom}_{\mathbf{X}}(M,N)$ *is seperated.*

PROOF. $\operatorname{Hom}_{\mathbf{X}}(-,-)$ is left exact on every basic open set by definition, and therefore left exact on arbitrary open sets since taking inverse limits preserves left exactness. The seperatedness follows from the preceding lemma. □

We associate to a presheaf F on $\mathbf{X}$ a sheaf $\widetilde{F}$ as follows. For any open set $\mathbf{U}$ let $\widetilde{F}(\mathbf{U})$ be the set of elements $(s_P) \in \prod_{P \in \mathbf{U}} F_P$ such that for every $P \in \mathbf{U}$ there is an open neighbourhood $P \in \mathbf{V} \subseteq \mathbf{U}$ and $t \in F(\mathbf{V})$ having the property $\rho_{\mathbf{V},Q}(t) = s_Q$ for every $Q \in \mathbf{V}$. Note that there is a functorial morphism $F \to \widetilde{F}$ of presheaves which is a monomorphism if F is seperated; it induces an isomorphism $F_P \simeq \widetilde{F}_P$ for every P in $\mathbf{X}$. We denote for every pair M, N of objects in $\mathcal{A}$ by $\widetilde{\operatorname{Hom}}_{\mathbf{X}}(M,N)$ the sheaf associated to $\operatorname{Hom}_{\mathbf{X}}(M,N)$; it is the *sheaf of local morphisms* from M to N. Some of the basic properties are as follows.

PROPOSITION 13.7. *The functor $\widetilde{\operatorname{Hom}}_{\mathbf{X}}(-,-)\colon \mathcal{A}^{\mathrm{op}} \times \mathcal{A} \longrightarrow \operatorname{Sh}\mathbf{X}$ has the following properties:*

(1) *$\widetilde{\operatorname{Hom}}_{\mathbf{X}}(-,-)$ is left exact in both arguments.*
(2) *$\widetilde{\operatorname{Hom}}_{\mathbf{X}}(M,-)$ commutes with direct limits if M is finitely presented.*
(3) *$\widetilde{\operatorname{Hom}}_{\mathbf{X}}(M,-)$ is faithful if M is a generator for $\mathcal{A}$.*

PROOF. (1) This follows from Proposition 13.6 since the sheafification functor $\operatorname{Presh}\mathbf{X} \to \operatorname{Sh}\mathbf{X}$ is exact.

(2) A morphism $F \to G$ of sheaves is an isomorphism if it induces isomorphisms $F_P \to G_P$ for every P in $\mathbf{X}$. Given a direct limit $N = \varinjlim N_i$ we have an induced isomorphism $\varinjlim \widetilde{\operatorname{Hom}}_{\mathbf{X}}(M,N_i)_P \to \widetilde{\operatorname{Hom}}_{\mathbf{X}}(M,\varinjlim N_i)_P$ for every P by Lemma 13.4 since M is finitely presented. Thus the assertion follows.

(3) Let $\phi\colon L \to N$ be a non-zero morphism in $\mathcal{A}$. If M is a generator for $\mathcal{A}$, then we find morphisms $\alpha\colon M \to L$ and $\beta\colon N \to P$ for some $P \in \mathbf{X}$ such that $\beta \circ \phi \circ \alpha \neq 0$. We obtain the following commutative diagram of canonical morphisms:

$$\begin{array}{ccccc}
\operatorname{Hom}_{\mathcal{A}}(M,L) & \xrightarrow{\operatorname{Hom}(M,\phi)} & \operatorname{Hom}_{\mathcal{A}}(M,N) & \xrightarrow{\operatorname{Hom}(M,\beta)} & \operatorname{Hom}_{\mathcal{A}}(M,P) \\
\downarrow & & \downarrow & & \downarrow \\
\widetilde{\operatorname{Hom}}_{\mathbf{X}}(M,L)_P & \xrightarrow{\widetilde{\operatorname{Hom}}(M,\phi)_P} & \widetilde{\operatorname{Hom}}_{\mathbf{X}}(M,N)_P & \xrightarrow{\widetilde{\operatorname{Hom}}(M,\beta)_P} & \widetilde{\operatorname{Hom}}_{\mathbf{X}}(M,P)_P \\
\downarrow & & \downarrow & & \downarrow \\
\operatorname{Hom}_{\mathcal{A}_P}(M,L) & \xrightarrow{\operatorname{Hom}(M,\phi)} & \operatorname{Hom}_{\mathcal{A}_P}(M,N) & \xrightarrow{\operatorname{Hom}(M,\beta)} & \operatorname{Hom}_{\mathcal{A}_P}(M,P)
\end{array}$$

The right hand column is an isomorphism according to the definition of $\mathcal{A}_P$, and the upper row is non-zero since $\beta \circ \phi \circ \alpha \neq 0$. Thus $\widetilde{\operatorname{Hom}}_{\mathbf{X}}(M,\phi) \neq 0$. □

The following example motivates our definition of the sheaf $\widetilde{\operatorname{Hom}}_{\mathbf{X}}(M,N)$; it is taken from [**26**, Chap. VI]. Let $(\mathbf{X}, \mathcal{O}_{\mathbf{X}})$ be a noetherian scheme and denote by $\operatorname{Qcoh}\mathbf{X}$ the category of quasi-coherent sheaves on $\mathbf{X}$. This is a locally coherent category and the coherent sheaves form the finitely presented objects in $\operatorname{Qcoh}\mathbf{X}$.

EXAMPLE 13.8. The Gabriel spectrum of $\operatorname{Qcoh}\mathbf{X}$ can be identified with $\mathbf{X}$ and for every pair M, N in $\operatorname{Qcoh}\mathbf{X}$ the sheaf of local morphisms $M \to N$ (which sends an open subset $\mathbf{U}$ of $\mathbf{X}$ to $\operatorname{Hom}(M|_{\mathbf{U}}, N|_{\mathbf{U}})$) is isomorphic to the presheaf $\operatorname{Hom}_{\mathbf{X}}(M,N)$; in particular $\operatorname{Hom}_{\mathbf{X}}(M,N)$ is a sheaf.

The proof can be reduced to the affine case, i.e. $(\mathbf{X}, \mathcal{O}_{\mathbf{X}}) = (\operatorname{Spec} R, \mathcal{O}_R)$ for some commutative noetherian ring R. This special case will be discussed in Theorem 15.9 and Theorem 15.11.

It is often useful to consider a relative version of the presheaf $\operatorname{Hom}_{\mathbf{X}}(M, N)$ for two objects M, N in $\mathcal{A}$. To this end we call a subset $\mathbf{Y} \subseteq \mathbf{X}$ *Ziegler-closed* if it is of the form $\mathbf{Y} = \{P \in \mathbf{X} \mid \operatorname{Hom}(\mathcal{S}, P) = 0\}$ for some Serre subcategory $\mathcal{S}$ of $\mathcal{C}$. The quotient category $\mathcal{A}/\mathcal{T}$ for $\mathcal{T} = \varinjlim \mathcal{S}$ is locally coherent and the section functor identifies the Gabriel spectrum of $\mathcal{A}/\mathcal{T}$ with $\mathbf{Y}$, see Proposition A.8. Note that $(\mathbf{Y} \cap \mathbf{U}_X)_{X \in \mathcal{C}}$ forms a basis of open subsets for $\mathbf{Y}$. Given two objects M, N in $\mathcal{A}$, let $\operatorname{Hom}_{\mathbf{Y}}(M, N)$ be the presheaf of Hom-groups which is defined as above on the Gabriel spectrum of $\mathcal{A}/\mathcal{T}$, and $\widetilde{\operatorname{Hom}}_{\mathbf{Y}}(M, N)$ denotes the associated sheaf. Thus

$$\operatorname{Hom}_{\mathbf{Y}}(M, N)(\mathbf{U}) = \varprojlim_{\mathbf{Y} \cap \mathbf{U}_X \subseteq \mathbf{U}} \operatorname{Hom}_{\mathcal{A}/(\mathcal{T} \cup \mathcal{T}_X)}(M, N)$$

for every open $\mathbf{U} \subseteq \mathbf{Y}$. Denoting by $i \colon \mathbf{Y} \to \mathbf{X}$ the inclusion we obtain the following commutative diagram of morphisms in $\operatorname{Presh} \mathbf{X}$

$$\begin{array}{ccc} \operatorname{Hom}_{\mathbf{X}}(M, N) & \longrightarrow & i_* \operatorname{Hom}_{\mathbf{Y}}(M, N) \\ \downarrow & & \downarrow \\ \widetilde{\operatorname{Hom}}_{\mathbf{X}}(M, N) & \longrightarrow & i_* \widetilde{\operatorname{Hom}}_{\mathbf{Y}}(M, N) \end{array}$$

where the upper morphism is induced by the collection of quotient functors $\mathcal{A}/\mathcal{T}_X \to \mathcal{A}/(\mathcal{T} \cup \mathcal{T}_X)$, $X \in \mathcal{C}$. This diagram gives rise to an alternative description of $\widetilde{\operatorname{Hom}}_{\mathbf{Y}}(M, N)$.

PROPOSITION 13.9. *There is a natural map* $\widetilde{\operatorname{Hom}}_{\mathbf{X}}(M, N)|_{\mathbf{Y}} \to \widetilde{\operatorname{Hom}}_{\mathbf{Y}}(M, N)$ *which is an isomorphism if M is finitely presented.*

PROOF. Restriction $\operatorname{Sh} \mathbf{X} \to \operatorname{Sh} \mathbf{Y}$, $M \mapsto M|_{\mathbf{Y}}$, is left adjoint to $i_* \colon \operatorname{Sh} \mathbf{Y} \to \operatorname{Sh} \mathbf{X}$. Thus $\widetilde{\operatorname{Hom}}_{\mathbf{X}}(M, N) \to i_* \widetilde{\operatorname{Hom}}_{\mathbf{Y}}(M, N)$ gives a map $\widetilde{\operatorname{Hom}}_{\mathbf{X}}(M, N)|_{\mathbf{Y}} \to \widetilde{\operatorname{Hom}}_{\mathbf{Y}}(M, N)$ which induces an isomorphism on each stalk by Lemma 13.4 whenever M is finitely presented. □

Our next aim is to show that certain functors between locally coherent Grothendieck categories preserve the (pre)sheaves of local morphisms. Let $f \colon \mathcal{A} \to B$ be an exact functor between locally coherent categories which preserves coproducts and sends finitely presented objects to finitely presented objects. We denote by $g \colon \mathcal{B} \to \mathcal{A}$ a right adjoint of f and let $\mathbf{X} = \operatorname{Sp} \mathcal{A}$ and $\mathbf{Y} = \operatorname{Sp} \mathcal{B}$.

LEMMA 13.10. *The assignment* $\mathbf{U} \mapsto \mathbf{U}^f = \bigcup_{\mathbf{U}_X \subseteq \mathbf{U}} \mathbf{U}_{f(X)}$ *induces an inclusion preserving map between the collections of open subsets in* $\mathbf{X}$ *and* $\mathbf{Y}$.

LEMMA 13.11. *Suppose that $g(M)$ is indecomposable for every $M \in \mathbf{Y}$. Then g induces a continuous map $\gamma \colon \mathbf{Y} \to \mathbf{X}$. Moreover, $\gamma^{-1}(\mathbf{U}) = \mathbf{U}^f$ for every open $\mathbf{U} \subseteq \mathbf{X}$.*

PROOF. Observe first that $g(M)$ is injective for every injective object M in $\mathcal{B}$ since f is exact. We obtain therefore a map $\mathbf{Y} \to \mathbf{X}$ if $g(M)$ is indecomposable for every $M \in \mathbf{Y}$. The adjointness isomorphism $\operatorname{Hom}_{\mathcal{A}}(X, g(M)) \simeq \operatorname{Hom}_{\mathcal{B}}(f(X), M)$ shows that $\gamma^{-1}(\mathbf{U}_X) = \mathbf{U}_{f(X)}$ for every finitely presented X in $\mathcal{A}$. The assertion now follows. □

We denote for every presheaf F on $\mathbf{Y}$ by $f_* F$ the presheaf on $\mathbf{X}$ which sends $\mathbf{U} \subseteq \mathbf{X}$ to $F(\mathbf{U}^f)$.

PROPOSITION 13.12. *Let $f\colon \mathcal{A} \to \mathcal{B}$ be an exact functor between locally coherent categories which preserves coproducts and sends finitely presented objects to finitely presented objects. Let M and N be objects in $\mathcal{A}$.*

(1) *The functor f induces a morphism $\operatorname{Hom}_{\mathbf{X}}(M,N) \to f_* \operatorname{Hom}_{\mathbf{Y}}(f(M),f(N))$ of presheaves on $\mathbf{X}$ which is functorial in M and N.*
(2) *Suppose that the right adjoint of f induces a map $\gamma\colon \mathbf{Y} \to \mathbf{X}$. Then $f_*\widetilde{F} = \gamma_*\widetilde{F}$ for every presheaf F on $\mathbf{Y}$, and f induces a morphism $\widetilde{\operatorname{Hom}_{\mathbf{X}}}(M,N) \to \gamma_* \widetilde{\operatorname{Hom}_{\mathbf{Y}}}(f(M),f(N))$ of sheaves on $\mathbf{X}$ which is functorial in M and N.*

PROOF. (1) Let X be a finitely presented object in $\mathcal{A}$. The functor f induces an exact functor $\mathcal{A}/\mathcal{T}_X \to \mathcal{B}/\mathcal{T}_{f(X)}$ which induces a map

$$\operatorname{Hom}_{\mathcal{A}/\mathcal{T}_X}(M,N) \longrightarrow \operatorname{Hom}_{\mathcal{B}/\mathcal{T}_{f(X)}}(f(M),f(N)).$$

By definition, we have

$$\operatorname{Hom}_{\mathbf{X}}(M,N)(\mathbf{U}_X) = \operatorname{Hom}_{\mathcal{A}/\mathcal{T}_X}(M,N)$$

and

$$\operatorname{Hom}_{\mathbf{Y}}(f(M),f(N))(\mathbf{U}_{f(X)}) = \operatorname{Hom}_{\mathcal{B}/\mathcal{T}_{f(X)}}(f(M),f(N)).$$

We obtain therefore a map

$$\operatorname{Hom}_{\mathbf{X}}(M,N)(\mathbf{U}) \to f_* \operatorname{Hom}_{\mathbf{Y}}(f(M),f(N))(\mathbf{U})$$

for every basic open subset $\mathbf{U}$ of $\mathbf{X}$, and this gives a morphism $\operatorname{Hom}_{\mathbf{X}}(M,N) \to f_* \operatorname{Hom}_{\mathbf{Y}}(f(M),f(N))$ of presheaves on $\mathbf{X}$.

(2) Apply the preceding lemma and use part (1). □

13.2. The Zariski topology

In this section we consider a topology on $\operatorname{Ind} R$ which is usually different from the Ziegler topology. We take the sets $\mathbf{U}_\phi$, ϕ a map in $\operatorname{mod} R$, as basic open sets and call a subset $\mathbf{U}$ of $\operatorname{Ind} R$ *Zariski-open* if $\mathbf{U} = \bigcup_{\phi\in\Phi} \mathbf{U}_\phi$ for some collection Φ of maps in $\operatorname{mod} R$.

LEMMA 13.13. *The Zariski-open sets of* $\operatorname{Ind} R$ *form a topology on* $\operatorname{Ind} R$. *The family $(\mathbf{U}_\phi)_{\phi\in\operatorname{mod} R}$ is a basis of open sets which is closed under finite intersections.*

It is possible to derive the Zariski topology from the Ziegler topology on $\operatorname{Ind} R$.

PROPOSITION 13.14. *Let $\mathbf{U} \subseteq \operatorname{Ind} R$ and $M \in \operatorname{Ind} R$. Then the following are equivalent:*

(1) *M belongs to every Zariski-closed subset $\mathbf{V} \subseteq \operatorname{Ind} R$ containing $\mathbf{U}$.*
(2) *M belongs to every Ziegler-open and quasi-compact subset $\mathbf{V} \subseteq \operatorname{Ind} R$ containing $\mathbf{U}$.*

PROOF. It has been shown in Proposition 2.10 that a subset $\mathbf{V} \subseteq \operatorname{Ind} R$ is Ziegler-open and quasi-compact if and only if $\mathbf{V} = \operatorname{Ind} R \setminus \mathbf{U}_\phi$ for some map ϕ. The characterization of the Zariski closure of $\mathbf{U}$ follows from this fact. □

Taking the Zariski topology on $\operatorname{Ind} R$, we obtain a homeomorphism

$$\operatorname{Ind} R \xrightarrow{\sim} \operatorname{Sp} D(R), \quad M \mapsto T_M$$

onto the Gabriel spectrum of the locally coherent category $D(R)$. We will often view this homeomorphism as an identification. Note that the homeomorphism $\operatorname{Ind} R \to \operatorname{Sp} D(R)$ explains the term "Zariski topology" since the topology on the

Gabriel spectrum of the locally coherent category $D(R)$ generalizes the Zariski topology [**26**, Chap. VI].

We have seen in Corollary 9.9 that a coherent functor $f\colon \operatorname{Mod} S \to \operatorname{Mod} R$ induces a continuous map $\operatorname{Ind} S \to \operatorname{Ind} R$ with respect to the Ziegler topology provided that $f(M)$ is indecomposable for every $M \in \operatorname{Ind} S$. We have an analogous result for the Zariski topology.

THEOREM 13.15. *Let $f\colon \operatorname{Mod} S \to \operatorname{Mod} R$ be a coherent functor. Suppose there are subsets $\mathbf{U} \subseteq \operatorname{Ind} R$ and $\mathbf{V} \subseteq \operatorname{Ind} S$ such that $f(M) \in \mathbf{U}$ for all $M \in \mathbf{V}$. Then the map $\mathbf{V} \to \mathbf{U}$, $M \mapsto f(M)$, is continuous with respect to the induced Zariski topologies.*

PROOF. It follows from Corollary 9.6 that for every map ϕ in $\operatorname{mod} R$

$$\{M \in \mathbf{V} \mid f(M) \in \mathbf{U}_\phi\} = \mathbf{V} \cap \mathbf{U}_\psi$$

for some map ψ in $\operatorname{mod} S$. Thus the preimage under $\mathbf{V} \to \mathbf{U}$ of every basic open subset of $\mathbf{U}$ is a basic open subset of $\mathbf{V}$. □

We shall frequently consider Ziegler-closed subsets $\mathbf{X} = \mathbf{U}_\Phi$ of $\operatorname{Ind} R$ together with the Zariski topology which is induced from the Zariski topology on $\operatorname{Ind} R$. Note that the family $(\mathbf{U}_{\Phi\cup\phi})_{\phi\in\operatorname{mod} R}$ forms a basis of open sets for this topology on $\mathbf{X}$ since $\mathbf{U}_{\Phi\cup\phi} = \mathbf{U}_\Phi \cap \mathbf{U}_\phi$.

EXAMPLE 13.16. Let M be a finitely presented indecomposable module over an artin algebra R. Then $\{M\}$ is Zariski-closed and Zariski-open. To see this let $\phi\colon M \to N$ be an almost split map. Then $\mathbf{U}_\phi = \operatorname{Ind} R \setminus \{M\}$ and therefore $\{M\}$ is Zariski-closed. The set $\{M\}$ is Ziegler-closed by Proposition 6.17, and therefore Ziegler-open and compact by Proposition 2.10. Thus $\{M\} = \mathbf{U}_\psi$ for some map ψ in $\operatorname{mod} R$ by Proposition 2.10, and it follows that $\{M\}$ is Zariski-open.

13.3. Structure sheaves

We fix a Ziegler-closed subset $\mathbf{X} = \mathbf{U}_\Phi$ of $\operatorname{Ind} R$ together with its Zariski topology. Our aim in this section is to construct from the rings of definable scalars over R a sheaf of rings $\mathcal{O}_\mathbf{X}$ on $\mathbf{X}$. This sheaf was first introduced by Prest [**63**]; it will be obtained from the *presheaf of definable scalars* $R_\mathbf{X}$ which is defined as follows. Let

$$R_\mathbf{X}(\mathbf{U}) = \varprojlim_{\mathbf{U}_{\Phi\cup\phi}\subseteq\mathbf{U}} R_{\Phi\cup\phi}$$

for every open $\mathbf{U} \subseteq \mathbf{X}$, where the structural morphisms $R_{\Phi\cup\phi} \to R_{\Phi\cup\psi}$ for every inclusion $\mathbf{U}_{\Phi\cup\psi} \subseteq \mathbf{U}_{\Phi\cup\phi}$ are taken from Proposition 11.13. We obtain restriction morphisms $\rho_{\mathbf{UV}}\colon R_\mathbf{X}(\mathbf{U}) \to R_\mathbf{X}(\mathbf{V})$ for every inclusion $\mathbf{V} \subseteq \mathbf{U}$ and have therefore constructed a presheaf of rings. Observe that the notation $R_\mathbf{X}$ for this presheaf is also used to denote the ring of definable scalars for $\mathbf{X}$; however, the meaning of the symbol $R_\mathbf{X}$ will always be clear from the context. We consider now the sheaf of rings $\mathcal{O}_\mathbf{X}$ which is associated to $R_\mathbf{X}$. Some basic properties of $\mathcal{O}_\mathbf{X}$ follow from our previous discussion in this chapter if we identify $\mathbf{X}$ with the corresponding Ziegler-closed subset of the Gabriel spectrum of $D(R)$. In order to state them let us recall that for every R-module M the ring of definable scalars R_{Φ_M} for $\Phi_M = \{\phi \in \operatorname{mod} R \mid M \text{ is } \phi\text{-injective}\}$ is denoted by R_M.

PROPOSITION 13.17. *The presheaf of definable scalars $R_\mathbf{X}$ on $\mathbf{X}$ has the following properties:*

(1) *$R_{\mathbf{X}} = \operatorname{End}_{\mathbf{X}}(T_R)$ and $\mathcal{O}_{\mathbf{X}} = \widetilde{\operatorname{End}}_{\mathbf{X}}(T_R)$.*
(2) *$R_{\mathbf{X}}$ is a seperated presheaf of rings.*
(3) *The canonical homomorphism $R_\Phi = \Gamma(\mathbf{X}, R_{\mathbf{X}}) \longrightarrow \Gamma(\mathbf{X}, \mathcal{O}_{\mathbf{X}})$ is injective.*
(4) *The stalk $(R_{\mathbf{X}})_P$ is isomorphic to R_P for every $P \in \mathbf{X}$.*

PROOF. (1) The original construction of the ring of definable scalars implies $R_{\Phi\cup\phi} = \operatorname{End}_{\mathbf{X}}(T_R)(\mathbf{U}_{\Phi\cup\phi})$ for every basic open subset $\mathbf{U}_{\Phi\cup\phi} \subseteq \mathbf{X}$. By definition, this carries over to arbitrary open subsets.

(2) follows from Proposition 13.6 and (1).

(3) follows from (2).

(4) follows from Lemma 13.4. □

Let us now specialize to the case $\mathbf{X} = \operatorname{Ind} R$. Thus we obtain

$$R_{\mathbf{U}} = \varprojlim_{\mathbf{U}_\phi \subseteq \mathbf{U}} R_\phi$$

for every Zariski-open subset $\mathbf{U}$ of $\operatorname{Ind} R$. The associated sheaf is denoted by $\mathcal{O}_R$. It is interesting to observe that $\mathcal{O}_{\mathbf{Y}} \simeq \mathcal{O}_R|_{\mathbf{Y}}$ for every Ziegler-closed subset $\mathbf{Y}$ of $\operatorname{Ind} R$ by Proposition 13.9. Let us now exhibit the functorial properties of the construction. To this end fix a ring homomorphism $f\colon R \to S$ and consider $\mathbf{Y} = \operatorname{Ind} S$ with the Zariski topology. We denote by $f_*\colon \operatorname{Mod} S \to \operatorname{Mod} R$ the corresponding restriction functor, and $\Phi \otimes_R S = \{\phi \otimes_R S \mid \phi \in \Phi\}$ for every collection Φ of maps in $\operatorname{mod} R$.

LEMMA 13.18. *If $\Phi \subseteq \Psi$ are collections of maps in* $\operatorname{mod} R$*, then f induces the following commutative diagram of ring homomorphisms:*

$$\begin{array}{ccc} R_\Phi & \longrightarrow & S_{\Phi\otimes_R S} \\ \downarrow & & \downarrow \\ R_\Psi & \longrightarrow & S_{\Psi\otimes_R S} \end{array}$$

PROOF. f extends to an exact functor $f'\colon C(R) \to C(S)$ which sends T_ϕ to $T_{\phi\otimes_R S}$ for every map ϕ in $\operatorname{mod} R$. Therefore f' induces for every Φ in $\operatorname{mod} R$ an exact functor $C(R)/\mathcal{S}_\Phi \to C(S)/\mathcal{S}_{\Phi\otimes_R S}$ which induces the map $R_\Phi \to S_{\Phi\otimes_R S}$. □

The next proposition shows that the construction of the presheaf $R_{\mathbf{X}}$ is in some weak sense functorial.

PROPOSITION 13.19. *Let $f\colon R \to S$ be a ring homomorphism.*
(1) *$\mathbf{U} \mapsto \mathbf{U}^f = \bigcup_{\mathbf{U}_\phi \subseteq \mathbf{U}} \mathbf{U}_{\phi\otimes_R S}$ induces an inclusion preserving map between the collections of Zariski-open sets in* $\operatorname{Ind} R$ *and* $\operatorname{Ind} S$.
(2) *f induces ring homomorphism $f_{\mathbf{U}}\colon R_{\mathbf{X}}(\mathbf{U}) \to S_{\mathbf{Y}}(\mathbf{U}^f)$ for every Zariski-open $\mathbf{U} \subseteq \operatorname{Ind} R$ which satisfy $f_{\mathbf{V}} \circ \rho_{\mathbf{UV}} = \rho_{\mathbf{U}^f\mathbf{V}^f} \circ f_{\mathbf{U}}$ for all $\mathbf{V} \subseteq \mathbf{U}$.*
(3) *f induces a ring homomorphism $R_{f_*(M)} \to S_M$ for every S-module M.*

PROOF. Apply the preceding lemma. □

By abuse of notation, we denote by $f_*S_{\mathbf{Y}}$ the presheaf on $\mathbf{X}$ which sends $\mathbf{U} \subseteq \mathbf{X}$ to $S_{\mathbf{U}^f}$ and reformulate the preceding proposition as follows.

COROLLARY 13.20. *A ring homomorphism $f\colon R \to S$ induces a homomorphism $R_{\mathbf{X}} \to f_*S_{\mathbf{Y}}$ of presheaves on $\mathbf{X}$.*

We obtain a better result if the restriction functor f_* preserves the indecomposability of indecomposable pure-injectives.

COROLLARY 13.21. *Let $f\colon R \to S$ be a ring homomorphism. Suppose there are Ziegler-closed subsets $\mathbf{X} \subseteq \operatorname{Ind} R$ and $\mathbf{Y} \subseteq \operatorname{Ind} S$ such that $f_*(M) \in \mathbf{X}$ for every $M \in \mathbf{Y}$, and consider the Zariski topologies on $\mathbf{X}$ and $\mathbf{Y}$. Then f induces a homomorphism $(\mathbf{Y}, \mathcal{O}_{\mathbf{Y}}) \to (\mathbf{X}, \mathcal{O}_{\mathbf{X}})$ of ringed spaces.*

PROOF. f_* induces a continuous map $\phi\colon \mathbf{Y} \to \mathbf{X}$ by Theorem 13.15. We obtain a homomorphism $R_{\mathbf{X}} \to f_* S_{\mathbf{Y}} = \phi_* S_{\mathbf{Y}}$ of presheaves which extends to a homomorphism $\mathcal{O}_{\mathbf{X}} \to \phi_* \mathcal{O}_{\mathbf{Y}}$. □

Replacing a ring S by the category $\operatorname{proj} S$ of finitely generated projective S-modules, we obtain from $\mathcal{O}_R$ a sheaf $\mathcal{O}_{\operatorname{proj} R}$ of additive categories if we define $\mathcal{O}_{\operatorname{proj} R}(\mathbf{U}) = \operatorname{proj} \mathcal{O}_R(\mathbf{U})$. We call a sheaf of skeletally small additive categories an *additive space* and it is clear how their morphisms are defined. This concept allows us to extend the functorial properties of the sheaf $\mathcal{O}_R$.

COROLLARY 13.22. *Let $f\colon \operatorname{Mod} R \to \operatorname{Mod} S$ be an exact functor which commutes with products and coproducts. Suppose there are Ziegler-closed subsets $\mathbf{X} \subseteq \operatorname{Ind} R$ and $\mathbf{Y} \subseteq \operatorname{Ind} S$ such that $f(M) \in \mathbf{X}$ for every $M \in \mathbf{Y}$, and consider the Zariski topologies on $\mathbf{X}$ and $\mathbf{Y}$. Then f induces a homomorphism $(\mathbf{Y}, \mathcal{O}_{\operatorname{proj} S}|_{\mathbf{Y}}) \to (\mathbf{X}, \mathcal{O}_{\operatorname{proj} R}|_{\mathbf{X}})$ of additive spaces.*

PROOF. f has a left adjoint which restricts to an additive functor $g\colon \operatorname{proj} R \to \operatorname{proj} S$. It is clear that $f = g_*$. Replacing ring homomorphisms $R' \to S'$ by additive functors $\operatorname{proj} R' \to \operatorname{proj} S'$, the proof of Corollary 13.21 carries over to the present situation. □

COROLLARY 13.23. *Let $f\colon \operatorname{Mod} S \to \operatorname{Mod} R$ be an exact functor which commutes with products and coproducts. Suppose also that $f(M)$ is indecomposable for every $M \in \operatorname{Ind} S$. Then the functor f induces a homomorphism $(\operatorname{Ind} S, \mathcal{O}_{\operatorname{proj} S}) \to (\operatorname{Ind} R, \mathcal{O}_{\operatorname{proj} R})$ of additive spaces.*

COROLLARY 13.24. *The additive space $(\operatorname{Ind} R, \mathcal{O}_{\operatorname{proj} R})$ is a Morita invariant of the ring R.*

It is easy to compute the ring of global sections for a ring of finite representation type.

PROPOSITION 13.25. *Let R be a ring of finite representation type and let $\mathbf{X} = \operatorname{Ind} R$. Then $\Gamma(\mathbf{X}, \mathcal{O}_R) \simeq \prod_{M \in \mathbf{X}} \operatorname{Biend}_R(M)$.*

PROOF. It follows from Theorem 6.24 that every $M \in \mathbf{X}$ is endofinite and therefore $\mathbf{X}$ carries the discrete topology, for instance by Proposition 6.17. Thus $\Gamma(\mathbf{X}, \mathcal{O}_R)$ is the product of the stalks $(\mathcal{O}_R)_M$. It follows from Proposition 13.17 and Corollary 11.10 that $(\mathcal{O}_R)_M \simeq \operatorname{Biend}_R(M)$ for every $M \in \mathbf{X}$. Thus $\Gamma(\mathbf{X}, \mathcal{O}_R) \simeq \prod_{M \in \mathbf{X}} \operatorname{Biend}_R(M)$. □

13.4. Associated sheaves of modules

We fix a Ziegler-closed subset $\mathbf{X} = \mathbf{U}_\Phi$ of $\operatorname{Ind} R$ together with its Zariski topology, and we wish to assign to each module M a sheaf $\widetilde{M_{\mathbf{X}}}$. To this end let $\mathcal{A} = D(R)/\mathcal{T}_\Phi$ and identify $\mathbf{X} = \operatorname{Sp} \mathcal{A}$. We shall apply the construction of the (pre)sheaf of local morphisms from the beginning of this chapter. To each module M we assign the presheaf $M_{\mathbf{X}} = \operatorname{Hom}_{\mathbf{X}}(T_R, T_M)$. Thus we have for every open $\mathbf{U} \subseteq \mathbf{X}$

$$M_{\mathbf{X}}(\mathbf{U}) = \varprojlim_{\mathbf{U}_{\Phi \cup \phi} \subseteq \mathbf{U}} \operatorname{Hom}_{D(R)/\mathcal{T}_{\Phi \cup \phi}}(T_R, T_M).$$

PROPOSITION 13.26. *$M_{\mathbf{X}}$ is a seperated presheaf.*

PROOF. Follows from Proposition 13.6. □

We denote by $\widetilde{M_{\mathbf{X}}}$ the sheaf associated to $M_{\mathbf{X}}$. If $\mathbf{X} = \operatorname{Ind} R$, then we write simply $\widetilde{M}$ instead of $\widetilde{M_{\mathbf{X}}}$. Note that for arbitrary $\mathbf{X}$ the sheaf $\widetilde{M_{\mathbf{X}}}$ can be obtained from the globally defined sheaf $\widetilde{M}$.

PROPOSITION 13.27. *$\widetilde{M_{\mathbf{X}}}$ is the restriction of $\widetilde{M}$ to $\mathbf{X}$, i.e. $\widetilde{M_{\mathbf{X}}} = \widetilde{M}|_{\mathbf{X}}$.*

PROOF. Follows from Proposition 13.9. □

We state now some basic properties of the sheaf associated to an R-module. To this end denote for every ringed space $\mathcal{O}$ by $\operatorname{Mod}\mathcal{O}$ the category of $\mathcal{O}$-modules.

PROPOSITION 13.28. *The assignment $M \mapsto \widetilde{M_{\mathbf{X}}}$ induces a functor $\operatorname{Mod} R \to \operatorname{Mod}\mathcal{O}_{\mathbf{X}}$. This functor sends pure left exact sequences to left exact sequences and preserves direct limits. It is faithful on the definable subcategory corresponding to $\mathbf{X}$. Moreover, it is faithful if $\mathbf{X}$ contains the indecomposable injective R-modules.*

PROOF. For every open $\mathbf{U} \subseteq \mathbf{X}$, the ring $R_{\mathbf{X}}(\mathbf{U})$ acts on $M_{\mathbf{X}}(\mathbf{U})$, and this action is compatible with the restriction maps corresponding to inclusions $\mathbf{V} \subseteq \mathbf{U}$. Therefore $\widetilde{M_{\mathbf{X}}}$ is an $\mathcal{O}_{\mathbf{X}}$-module. The functor $M \mapsto \widetilde{M_{\mathbf{X}}}$ is left exact on pure-exact sequences and preserves direct limits by Proposition 13.7. Adapting the argument in the proof of (3) in Proposition 13.7, one shows that the functor is faithful under suitable assumptions. □

Taking global sections induces for every R-module M a functorial morphism

$$\gamma_M \colon M \simeq \operatorname{Hom}_{D(R)}(T_R, T_M) \to \operatorname{Hom}_{D(R)/\mathcal{T}_\Phi}(T_R, T_M) \simeq \Gamma(\mathbf{X}, M_{\mathbf{X}}) \to \Gamma(\mathbf{X}, \widetilde{M_{\mathbf{X}}}).$$

The group $\Gamma(\mathbf{X}, \widetilde{M_{\mathbf{X}}})$ becomes an R-module via the ring homomorphism $R \to \Gamma(\mathbf{X}, \mathcal{O}_{\mathbf{X}})$, and it is easily checked that γ_M is R-linear.

PROPOSITION 13.29. *The map $\gamma_M \colon M \to \Gamma(\mathbf{X}, \widetilde{M_{\mathbf{X}}})$ is injective if M belongs to the definable subcategory corresponding to $\mathbf{X}$, or if $\mathbf{X}$ contains the indecomposable injective R-modules.*

PROOF. If M belongs to the definable subcategory corresponding to $\mathbf{X}$, then the map $\operatorname{Hom}_{D(R)}(T_R, T_M) \to \operatorname{Hom}_{D(R)/\mathcal{T}_\Phi}(T_R, T_M)$ is an isomorphism. Thus γ_M is injective since $M_{\mathbf{X}}$ is seperated. If $\mathbf{X}$ contains the injectives, then for every $0 \neq m \in M$ there is $P \in \mathbf{X}$ and $\phi \in \operatorname{Hom}_R(M, P)$ whith $\phi(m) \neq 0$. Using Lemma 13.4 we obtain the following commutative diagram

$$\begin{array}{ccccc}
M & \xrightarrow{\sim} & \operatorname{Hom}_{D(R)}(T_R, T_M) & \xrightarrow{\operatorname{Hom}(T_R, T_\phi)} & \operatorname{Hom}_{D(R)}(T_R, T_P) \\
\downarrow{\scriptstyle \alpha} & & \downarrow & & \downarrow{\scriptstyle \wr} \\
(M_{\mathbf{X}})_P & \longrightarrow & \operatorname{Hom}_{D(R)_P}(T_R, T_M) & \longrightarrow & \operatorname{Hom}_{D(R)_P}(T_R, T_P)
\end{array}$$

and $\phi(m) \neq 0$ implies $\alpha(m) \neq 0$. Therefore $\gamma_M(m) \neq 0$ since α is the composition of γ_M with the canonical map $\Gamma(\mathbf{X}, \widetilde{M_{\mathbf{X}}}) \to (M_{\mathbf{X}})_P$. □

CHAPTER 14

Tame hereditary algebras

In this chapter we investigate the sheaf of definable scalars for a tame hereditary artin algebra R. Restricting to an appropriate Ziegler-closed subset $\mathbf{X}$ of $\operatorname{Ind} R$ it is shown that the sheaf $\mathcal{O}_{\mathbf{X}}$ yields a curve which parametrizes the finitely generated indecomposable R-modules. This curve has been studied by Baer, Geigle, and Lenzing [**57, 9**], and its centre by Crawley-Boevey [**16**]. The reinterpretation of this parametrizing curve via rings of definable scalars was suggested by Prest [**65**]. In fact, the approach of Prest yields a different curve $\mathbf{Y}$ which admits an involution δ (induced by the duality $\operatorname{Hom}_k(-,k)$ where k denotes the centre of R, leaving fixed only the unique generic point of $\mathbf{Y}$) such that $\mathbf{X} = \mathbf{Y}/\delta$.

14.1. Preliminaries

In this section we study the functor $\operatorname{Mod} R \to D(R)/\mathcal{T}$, $M \mapsto T_M$, where $\mathcal{T}$ denotes the finite type localizing subcategory corresponding to some Ziegler-closed subset $\mathbf{X}$ of $\operatorname{Ind} R$. Throughout this section R denotes an artin k-algebra over a commutative artinian ring k. The usual k-duality sends $M \in \operatorname{Mod} k$ to $M^* = \operatorname{Hom}_k(M, E)$ where E denotes an injective envelope of $k/\operatorname{rad} k$. We begin our discussion with some preliminary results. Let $\mathbf{C} \subseteq \operatorname{Ind} R$ be a set of finitely generated indecomposables. We denote by $\mathbf{X}_0 = \overline{\mathbf{C}}$ the Ziegler closure of $\mathbf{C}$ and $\mathbf{X} = \overline{\mathbf{C}} \backslash \mathbf{C}$ denotes the Ziegler-closed subset of all non-finitely generated indecomposables belonging to the Ziegler closure of $\mathbf{C}$. Recall from Theorem that $\mathbf{X}$ determines a localizing subcategory $\mathcal{T}$ of $D(R)$; a functor F in $D(R)$ belongs to $\mathcal{T}$ if and only if $\operatorname{Hom}(F, T_M) = 0$ for all M in $\mathbf{X}$. We are interested in a description of the quotient $D(R)/\mathcal{T}$. To this end we denote by $\mathcal{C} = \operatorname{add} \mathbf{C}$ the full subcategory of $\operatorname{mod} R$ which is formed by all finite coproducts of modules in $\mathbf{C}$.

LEMMA 14.1. *Suppose that $\mathcal{C}$ is contravariantly finite in* $\operatorname{mod} R$. *Consider the functor*

$$p\colon D(R) = (\operatorname{mod} R^{\mathrm{op}}, \mathrm{Ab}) \longrightarrow (\mathcal{C}^*, \mathrm{Ab}) = \mathcal{A}, \quad F \mapsto F|_{\mathcal{C}^*}$$

and the quotient functor $q\colon \mathcal{A} \to \mathcal{A}/\mathcal{A}_0$ where $\mathcal{A}_0$ denotes the localizing subcategory generated by all finite length objects in $\mathcal{A}$.

(1) *$\mathcal{T}_0 = \operatorname{Ker} p$ is the localizing subcategory corresponding to $\mathbf{X}_0$, and p induces an equivalence between $D(R)/\mathcal{T}_0$ and $\mathcal{A}$.*
(2) *$\mathcal{T} = \operatorname{Ker}(q \circ p)$ is the localizing subcategory corresponding to $\mathbf{X}$, and $q \circ p$ induces an equivalence between $D(R)/\mathcal{T}$ and $\mathcal{A}/\mathcal{A}_0$.*

PROOF. (1) Denote by $\mathcal{S}_0$ the Serre subcategory of all functors in $C(R)$ vanishing on $\mathcal{C}^*$. It follows from the isomorphism $M \otimes_R X^* \simeq \operatorname{Hom}_R(M, X)^*$ for M and X in $\operatorname{mod} R$ that for F in $C(R)$ we have $\operatorname{Hom}(F, T_M) = 0$ for every $M \in \mathbf{C}$ if and only if $F \in \mathcal{S}_0$. Therefore $\varinjlim \mathcal{S}_0$ is the localizing subcategory corresponding to $\mathbf{X}_0$ by Corollary 2.13. The assumption on $\mathcal{C}$ to be contravariantly finite in $\operatorname{mod} R$

implies that $\varinjlim \mathcal{S}_0$ is the category of all functors in $D(R)$ vanishing on $\mathcal{C}^*$, i.e. $\operatorname{Ker} p$, by [**54**, Theorem 3.4].

(2) The assignment $X \mapsto S_X = \operatorname{Hom}_{R^{\mathrm{op}}}(X^*, -)/\operatorname{rad}\operatorname{Hom}_{R^{\mathrm{op}}}(X^*, -)$ induces a bijection between $\mathbf{C}$ and the simple objects in $\mathcal{A}$. The injective envelope of S_X is T_X. Given $M \in \mathbf{X}_0$, the quotient functor p induces an isomorphism $\operatorname{Hom}_{D(R)}(F, T_M) \to \operatorname{Hom}_{\mathcal{A}}(F, T_M)$ since T_M is $\mathcal{T}_0$-closed. It follows that every non-zero morphism $S_X \to T_M$ induces a split monomorphism $X \to M$. Thus $\operatorname{Hom}(\mathcal{A}_0, T_M) = 0$ if and only if $M \in \mathbf{X}$, and therefore $\operatorname{Ker}(q \circ p)$ is the localizing subcategory corresponding to $\mathbf{X}$. □

We proceed with a discussion of the functor $\operatorname{Mod} R \to D(R)/\mathcal{T}$, $M \mapsto T_M$. If we impose some additional conditions on $\mathcal{C}$, then this functor is fully faithful on the full subcategory of R-modules having no direct summand in $\mathbf{C}$.

PROPOSITION 14.2. *Let $\mathbf{C} \subseteq \operatorname{Ind} R$ be a set of finitely generated indecomposables and denote by $\mathcal{T}$ the localizing subcategory corresponding to $\overline{\mathbf{C}} \setminus \mathbf{C}$. Suppose that $\mathcal{C} = \operatorname{add} \mathbf{C}$ has the following properties:*

(1) *$\mathcal{C}$ is contravariantly finite in $\operatorname{mod} R$ and contains the injectives;*
(2) *$\mathcal{C}$ contains a right almost split map $Y \to X$ for every $X \in \mathbf{C}$.*

Then the functor $\operatorname{Mod} R \to D(R)/\mathcal{T}$, $M \mapsto T_M$, induces an isomorphism

$$\operatorname{Hom}_R(M, N) \longrightarrow \operatorname{Hom}_{D(R)/\mathcal{T}}(T_M, T_N)$$

provided that N has no direct summand in $\mathbf{C}$. If $0 \to L \to M \to N \to 0$ is an exact sequence of R-modules with $\operatorname{Ext}^1_R(N, \mathcal{C}) = 0$, then $0 \to T_L \to T_M \to T_N \to 0$ is exact.

PROOF. We keep the notation from the preceding lemma. In addition we use the left adjoint $l\colon \mathcal{A} \to D(R)$ of the restriction functor $p\colon D(R) \to \mathcal{A}$. Note that $l(\operatorname{Hom}_{\mathcal{C}^*}(M, -)) = \operatorname{Hom}_{R^{\mathrm{op}}}(M, -)$ for every M in $\mathcal{C}^*$, and $l(T_N|_{\mathcal{C}^*}) = T_N$ for every R-module N. Also, $l \circ p(S_X) = S_X$ for each simple functor S_X corresponding to X in $\mathbf{C}$ since for any right almost split map $Y \to X$ in $\mathcal{C}$ the exact sequence

$$\operatorname{Hom}_{\mathcal{C}^*}(Y^*, -) \longrightarrow \operatorname{Hom}_{\mathcal{C}^*}(X^*, -) \longrightarrow S_X \to 0$$

is sent by l to the exact sequence

$$\operatorname{Hom}_{R^{\mathrm{op}}}(Y^*, -) \longrightarrow \operatorname{Hom}_{R^{\mathrm{op}}}(X^*, -) \longrightarrow S_X \to 0.$$

It is clear that p induces a bijection $\operatorname{Hom}_R(M, N) \to \operatorname{Hom}_{\mathcal{A}}(T_M, T_N)$ for every pair M, N in $\operatorname{Mod} R$ since $R^* \in \mathcal{C}$. Let us show that the map $\operatorname{Hom}_{\mathcal{A}}(F, T_N) \to \operatorname{Hom}_{\mathcal{A}/\mathcal{A}_0}(F, T_N)$ induced by q is bijective for every F in $\mathcal{A}$. If $q(\phi) = 0$ for some $\phi \in \operatorname{Hom}_{\mathcal{A}}(F, T_N)$, then $\operatorname{Im}\phi \in \mathcal{A}_0$. Assuming $\operatorname{Im}\phi \neq 0$ we obtain a non-zero morphism $\psi\colon S_X \to T_N$ in $\mathcal{A}$ for some X in $\mathbf{C}$. The morphism $l(\psi)$ extends to to a split monomorphism $T_X \to T_N$ in $D(R)$ which contradicts our assumption on N. It follows that $\phi = 0$. An element in $\operatorname{Hom}_{\mathcal{A}/\mathcal{A}_0}(F, T_N)$ is of the form $q(\phi)$ for some morphism $\phi\colon F' \to T_N/G$ with $F' \subseteq F$ and $G \subseteq T_N$ such that $F/F', G \in \mathcal{A}_0$. We have already seen that $G = 0$ and we need to extend ϕ to F. To this end observe that l sends F/F' to an object in $\varinjlim \mathcal{S}$ where $\mathcal{S}$ denotes the Serre subcategory of $C(R)$ generated by the simples S_X, $X \in \mathbf{C}$. Also, (ii) implies that $0 \to l(F') \to l(F) \to l(F/F') \to 0$ is exact. Therefore $l(\phi)$ extends to a morphism $\phi'\colon l(F) \to T_N$ since $\operatorname{Hom}(\mathcal{S}, T_N) = 0$ implies $\operatorname{Ext}^1(\varinjlim \mathcal{S}, T_N) = 0$ by Proposition A.6. Clearly, $q(p(\phi')) = q(\phi)$ and therefore $\operatorname{Hom}_{\mathcal{A}}(F, T_N) \to \operatorname{Hom}_{\mathcal{A}/\mathcal{A}_0}(F, T_N)$ is also surjective.

It remains to check the exactness property of $q \circ p$. Any exact sequence of R-modules $0 \to L \to M \to N \to 0$ induces an exact sequence

$$\operatorname{Tor}_1^R(N, X) \longrightarrow L \otimes_R X \longrightarrow M \otimes_R X \longrightarrow N \otimes_R X \longrightarrow 0$$

for every X in $\operatorname{mod} R^{\mathrm{op}}$. Combining the isomorphism

$$\operatorname{Ext}_R^1(N, (\operatorname{Tr} Y)^*) \simeq \operatorname{Tor}_1^R(N, \operatorname{Tr} Y)^*$$

from [**5**, Proposition I.3.3] with the assumption $\operatorname{Ext}_R^1(N, \mathcal{C}) = 0$ we obtain the exactness of

$$0 \longrightarrow T_L|_{\mathcal{C}^*} \longrightarrow T_M|_{\mathcal{C}^*} \longrightarrow T_N|_{\mathcal{C}^*} \longrightarrow 0.$$

Thus $0 \to T_L \to T_M \to T_N \to 0$ is exact in $D(R)/\mathcal{T}$. □

Let Γ be a component of the AR-quiver of R (not necessarily connected) such that

(Γ1) Γ contains the indecomposable injectives;
(Γ2) every point in Γ has only finitely many successors.

Identifying Γ with the corresponding points in $\operatorname{Ind} R$, we denote by $\mathbf{X} = \overline{\Gamma} \setminus \Gamma$ the Ziegler-closed subset of all non-finitely generated indecomposables belonging to the Ziegler closure of Γ. Let $\mathcal{T}_{\mathbf{X}}$ be the corresponding localizing subcategory of $D(R)$, i.e. F in $D(R)$ belongs to $\mathcal{T}_{\mathbf{X}}$ if and only if $\operatorname{Hom}(F, T_M) = 0$ for all $M \in \mathbf{X}$.

Proposition 14.3. *Let $\mathcal{C} = \operatorname{add} \Gamma$. Then there is a torsion theory $(\varinjlim \mathcal{C}, \mathcal{F})$ for* $\operatorname{Mod} R$. *Any R-module M has the following properties:*

(1) *$M \in \varinjlim \mathcal{C}$ if and only if M is a coproduct of modules in Γ.*
(2) *$M \in \mathcal{F}$ if and only if M has no direct summand in Γ.*

Proof. We need only the condition (Γ2). In fact this assumption implies that the functor $\operatorname{Hom}_R(X, -)$ is of finite length in $(\operatorname{mod} R, \mathrm{Ab})$ for every X in Γ. Also, every indecomposable X in $\operatorname{mod} R$ with $\operatorname{Hom}_R(\mathcal{C}, X) \neq 0$ belongs to $\mathcal{C}$. Thus $\mathcal{C}$ is closed under factor modules and extensions. We obtain a torsion theory $(\mathcal{C}, \mathcal{D})$ for $\operatorname{mod} R$ which extends to a torsion theory $(\varinjlim \mathcal{C}, \varinjlim \mathcal{D})$ by [**20**, Lemma 4.4]. The Auslander-Reiten formula implies $\operatorname{Ext}_R^1(\mathcal{D}, \mathcal{C}) = 0$ since $\operatorname{Tr}(\mathcal{C}^*) \subseteq \mathcal{C}$. We denote for every R-module M by cM the maximal submodule in $\varinjlim \mathcal{C}$, and we conclude from $\operatorname{Ext}_R^1(\mathcal{D}, \mathcal{C}) = 0$ that the inclusion $cM \to M$ is pure. The finite length of $\operatorname{Hom}_R(X, -)$ for every X in Γ implies that every module in $\varinjlim \mathcal{C}$ decomposes into a coproduct of modules in Γ; this follows, for instance, from [**20**, Theorem 3.2]. This is (1), and (2) follows immediately since every pure mono $X \to M$ with X in Γ splits. □

Theorem 14.4. *The functors*

$$\operatorname{Mod} R \longrightarrow D(R)/\mathcal{T}_{\mathbf{X}},\ M \mapsto T_M \quad \textit{and} \quad D(R)/\mathcal{T}_{\mathbf{X}} \longrightarrow \operatorname{Mod} R,\ F \mapsto \operatorname{Hom}(T_R, F)$$

induce mutually inverse equivalences between

(1) *the full subcategory of R-modules having no direct summand in Γ, and*
(2) *the full subcategory of objects F having a presentation*

$$(T_R)^{(I)} \longrightarrow (T_R)^{(J)} \longrightarrow F \longrightarrow 0.$$

The kernel of $\operatorname{Mod} R \to D(R)/\mathcal{T}_{\mathbf{X}}$ *consists of all maps between R-modules which factor through a coproduct of modules in Γ.*

PROOF. We denote by $\mathcal{F}$ the full subcategory of R-modules having no direct summand in Γ and denote by f the functor $\operatorname{Mod} R \to D(R)/\mathcal{T}_{\mathbf{X}}$. We can apply Proposition 14.2 since (Γ1) - (Γ2) imply the assumptions (1) - (2) in Proposition 14.2. Therefore f is fully faithful on $\mathcal{F}$. In order to describe the kernel of f consider a map $\phi\colon M \to N$ in $\operatorname{Mod} R$. We apply the preceding proposition and use the torsion functor $c\colon \operatorname{Mod} R \to \operatorname{Mod} R$ corresponding to $\varinjlim \mathcal{C}$. Thus we obtain the following commutative diagram with exact rows:

$$\begin{array}{ccccccccc} 0 & \longrightarrow & cM & \longrightarrow & M & \longrightarrow & M/cM & \longrightarrow & 0 \\ & & \downarrow{\scriptstyle \phi'} & & \downarrow{\scriptstyle \phi} & & \downarrow{\scriptstyle \phi''} & & \\ 0 & \longrightarrow & cN & \longrightarrow & N & \longrightarrow & N/cN & \longrightarrow & 0 \end{array}$$

If $f(\phi) = 0$, then $f(\phi'') = 0$ and therefore $\phi'' = 0$ since M/cM and N/cN belong to $\mathcal{F}$. Thus ϕ factors through cN which is a coproduct of modules in Γ. Conversely, if M is a coproduct of modules in Γ, we need to show that $f(M) = 0$. The isomorphism $M \otimes_R X^* \simeq \operatorname{Hom}_R(M, X)^*$ for X in $\operatorname{mod} R$ in combination with (Γ2) implies that T_M has finite length in $(\mathcal{C}^*, \mathrm{Ab})$ if $M \in \Gamma$. Therefore $f(M) = q \circ p(T_M) = 0$.

Now we want to show that for every F in $D(R)/\mathcal{T}_{\mathbf{X}}$ we have a presentation

$$(T_R)^{(I)} \longrightarrow (T_R)^{(J)} \longrightarrow F \longrightarrow 0$$

if and only if $F \simeq T_M$ for some $M \in \mathcal{F}$. One direction is clear since f is right exact and preserves coproducts. Suppose therefore that F has a presentation as above. Decomposing $R = cR \coprod R/cR$, we find a map $\psi\colon (R/cR)^{(I)} \to (R/cR)^{(J)}$ with $f(\psi) = \phi$ since $f(cR) = 0$ and $R/cR \in \mathcal{F}$. We have therefore $f(M) \simeq F$ for $M = \operatorname{Coker} \psi$. Also, $cM = 0$ and this implies that $M \in \mathcal{F}$.

It remains to show that $F \mapsto \operatorname{Hom}(T_R, F)$ is an inverse for $f|_{\mathcal{F}}$. Note that the R-action on $\operatorname{Hom}(T_R, F)$ is induced by the ring homomorphism $R \to \operatorname{End}(T_R)$. If $M \in \mathcal{F}$, then

$$M \simeq \operatorname{Hom}_R(R, M) \simeq \operatorname{Hom}_R(R/cR, M) \simeq \operatorname{Hom}(T_{R/cR}, T_M) \simeq \operatorname{Hom}(T_R, T_M)$$

and this proves the claim. □

14.2. A parametrizing curve

We assume in this section that R is a tame hereditary artin algebra. For definitions and the basic theory of tame hereditary algebras we refer to [**22, 70, 32**]. Let $\mathbf{I} \subseteq \operatorname{Ind} R$ be the set of finitely generated preinjective indecomposables, and denote by $\mathbf{X} = \bar{\mathbf{I}} \setminus \mathbf{I}$ the Ziegler-closed set of all non-finitely generated indecomposables belonging to the Ziegler closure of $\mathbf{I}$. We consider the Zariski topology on $\mathbf{X}$. The main result of this section is the following.

THEOREM 14.5. *The presheaf $R_{\mathbf{X}}$ of definable scalars on $\mathbf{X}$ is a sheaf, i.e. $R_{\mathbf{X}} = \mathcal{O}_{\mathbf{X}}$. It has the following properties:*

(1) $\Gamma(\mathbf{X}, \mathcal{O}_{\mathbf{X}}) \simeq R$.
(2) *The space $\mathbf{X}$ has dimension 1 and has a unique generic point G which is the unique generic R-module.*
(3) *Given $P \in \mathbf{X}$, the endomorphism ring $\operatorname{End}_R(P)$ is uniserial. Choosing a generator π_P for the radical of $\operatorname{End}_R(P)$, the map $P \mapsto \operatorname{Ker} \pi_P$ induces a bijection between $\mathbf{X} \setminus \{G\}$ and the isomorphism classes of simple regular R-modules.*

For the proof of this theorem we combine Geigle's analysis of the category $C(R)$ with the previous results in this chapter. Let us recall from [**32**] some properties of $C(R)$. To this end denote by $\operatorname{reg} R$ the full subcategory of regular modules in $\operatorname{mod} R$, and G denotes the unique generic module over R. Let $\mathcal{S}_{\mathbf{X}}$ be the Serre subcategory of functors F in $C(R)$ satisfying $\operatorname{Hom}(F, T_M) = 0$ for all $M \in \mathbf{X}$. Analogously, $\mathcal{S}_G$ is defined.

LEMMA 14.6. *The following holds:*

(1) *The functor* $\operatorname{reg} R \to C(R)/\mathcal{S}_{\mathbf{X}}$, $M \mapsto T_M$, *induces an equivalence between* $\operatorname{reg} R$ *and the category* $\mathcal{S}_0$ *of finite length objects in* $C(R)/\mathcal{S}_{\mathbf{X}}$.
(2) $\mathcal{S}_G$ *contains* $\mathcal{S}_{\mathbf{X}}$ *and* $\mathcal{S}_G/\mathcal{S}_{\mathbf{X}}$ *is equivalent to* $\mathcal{S}_0$.
(3) $\dim C(R)/\mathcal{S}_{\mathbf{X}} = 1$.

PROOF. (1) The functor is fully faithful by Proposition 14.2. It is dense by [**32**, Theorem 3.7].

(2) It has been shown in Proposition 6.23 that the map $M \mapsto \mathcal{S}_M$ induces a bijection between the generic R-modules and the Serre subcategories $\mathcal{S}$ of $C(R)$ containing all finite length objects such that $C(R)/\mathcal{S}$ is a length category with a unique simple object. In [**32**, Theorem 4.6] it is shown that the quotient category $C(R)/\mathcal{S}_{\mathbf{X}}$ modulo $\mathcal{S}_0$ is a length category with a unique simple object. Combining these facts, (2) follows.

(3) follows from the fact that the quotient category $C(R)/\mathcal{S}_{\mathbf{X}}$ modulo $\mathcal{S}_0$ is a length category. □

We shall also need the following technical lemma.

LEMMA 14.7. *Let $\mathcal{A}$ be a locally coherent category with Gabriel spectrum $\mathbf{Y} = \operatorname{Sp} \mathcal{A}$. Suppose that $\mathbf{U}_X = \emptyset$ for every finitely presented object X in $\mathcal{A}$ which is not of finite length. Then* $\operatorname{Hom}_{\mathbf{Y}}(M, N)$ *is a sheaf for every pair M, N in $\mathcal{A}$ such that N has no finite length subobject.*

PROOF. Let $F = \operatorname{Hom}_{\mathbf{Y}}(M, N)$. The assumptions on $\mathbf{Y}$ imply that every closed subset is either finite or coincides with $\mathbf{Y}$. Thus $\mathbf{Y}$ is a noetherian space. Therefore we need to check for two open subsets $\mathbf{U}_1, \mathbf{U}_2$ the following conditions:

(1) Let $\alpha \in F(\mathbf{U}_1 \cup \mathbf{U}_2)$. If $\rho_{\mathbf{U}_1 \cup \mathbf{U}_2 \mathbf{U}_1}(\alpha) = 0 = \rho_{\mathbf{U}_1 \cup \mathbf{U}_2 \mathbf{U}_2}(\alpha)$, then $\alpha = 0$.
(2) Let $\beta_1 \in F(\mathbf{U}_1)$ and $\beta_2 \in F(\mathbf{U}_2)$. If $\rho_{\mathbf{U}_1 \mathbf{U}_1 \cap \mathbf{U}_2}(\beta_1) = \rho_{\mathbf{U}_2 \mathbf{U}_1 \cap \mathbf{U}_2}(\beta_2)$, then $\rho_{\mathbf{U}_1 \cup \mathbf{U}_2 \mathbf{U}_1}(\alpha) = \beta_1$ and $\rho_{\mathbf{U}_1 \cup \mathbf{U}_2 \mathbf{U}_2}(\alpha) = \beta_2$ for some $\alpha \in F(\mathbf{U}_1 \cup \mathbf{U}_2)$.

Condition (1) holds since the presheaf F is seperated by Proposition 13.6. To check (2) we may exclude the case that $\mathbf{U}_1 = \emptyset$ or $\mathbf{U}_2 = \emptyset$. Thus there are finite length objects X_1, X_2 in $\mathcal{A}$ such that $\mathbf{U}_1 = \mathbf{U}_{X_1}$ and $\mathbf{U}_2 = \mathbf{U}_{X_2}$. Passing to a suitable quotient of $\mathcal{A}$ we may also assume that $\mathbf{Y} = \mathbf{U}_1 \cup \mathbf{U}_2$. Using the assumption that N has no proper finite length subobjects we find for each i subobjects $M_i \subseteq M$ with $M/M_i \in \mathcal{T}_{X_i}$ and $\alpha_i \in \operatorname{Hom}_{\mathcal{A}}(M_i, N)$ such that β_i is the image of α_i under the map

$$\operatorname{Hom}_{\mathcal{A}}(M_i, N) \longrightarrow \operatorname{Hom}_{\mathcal{A}/\mathcal{T}_{X_i}}(M, N) = F(\mathbf{U}_i).$$

If $\rho_{\mathbf{U}_1 \mathbf{U}_1 \cap \mathbf{U}_2}(\beta_1) = \rho_{\mathbf{U}_2 \mathbf{U}_1 \cap \mathbf{U}_2}(\beta_2)$, then $\alpha_1|_{M_1 \cap M_2} = \alpha_2|_{M_1 \cap M_2}$. We have $M = M_1 + M_2$ since $\mathbf{Y} = \mathbf{U}_1 \cup \mathbf{U}_2$, and therefore there exists $\alpha \in \operatorname{Hom}_{\mathcal{A}}(M, N) = F(\mathbf{U}_1 \cup \mathbf{U}_2)$ such that $\alpha_i = \alpha|_{M_i}$ for each i. This completes the proof of (2) and therefore F is a sheaf. □

PROOF OF THEOREM 14.5. We prove first that $R_{\mathbf{X}}$ is a sheaf. Using the bijection between Serre subcategories of $C(R)/\mathcal{S}_{\mathbf{X}}$ and Ziegler-closed subsets of $\mathbf{X}$, it follows from Lemma 14.6 that every non-empty open subset of $\mathbf{X}$ contains G. If $X \in C(R)/\mathcal{S}_{\mathbf{X}}$ is an object which is not of finite length, then $X \notin \mathcal{S}_G/\mathcal{S}_{\mathbf{X}}$ by Lemma 14.6. Thus $G \notin \mathbf{U}_X$ and $\mathbf{U}_X = \emptyset$ follows. We claim that T_R has no finite length subobject in $D(R)/\mathcal{T}_{\mathbf{X}}$. This follows from Theorem 14.4 and Lemma 14.6 since $\operatorname{Hom}_R(\operatorname{reg} R, R) = 0$. Thus we can apply Lemma 14.7 and $R_{\mathbf{X}} = \operatorname{End}_{\mathbf{X}}(T_R)$ is a sheaf.

(1) Theorem 14.4 implies that $\operatorname{End}_{D(R)/\mathcal{T}_{\mathbf{X}}}(T_R) \simeq R$ and therefore $\Gamma(\mathbf{X}, \mathcal{O}_{\mathbf{X}}) \simeq R$ since $R_{\mathbf{X}} = \mathcal{O}_{\mathbf{X}}$.

(2) The quotient category of $C(R)/\mathcal{S}_{\mathbf{X}}$ modulo all finite length objects is equivalent to $C(R)/\mathcal{S}_G$ by Lemma 14.6 and is therefore a length category. Using the bijection between Serre subcategories of $C(R)/\mathcal{S}_{\mathbf{X}}$ and Ziegler-closed subsets of $\mathbf{X}$, the assertion follows.

(3) The Grothendieck category $D(R)/\mathcal{T}_{\mathbf{X}}$ has Krull dimension 1. This follows form part (3) in Lemma 14.6 and Lemma B.5. Therefore Gabriel's analysis in [**26**] implies that taking injective envelopes induces a bijection between the simple objects in $C(R)/\mathcal{S}_{\mathbf{X}}$ and $C(R)/\mathcal{S}_G$ and the points in $\mathbf{X} = \operatorname{Sp}(D(R)/\mathcal{T}_{\mathbf{X}})$. The unique simple object in $C(R)/\mathcal{S}_G$ corresponds to G, and Lemma 14.6 implies that the simple objects in $C(R)/\mathcal{S}_{\mathbf{X}}$, and therefore the points in $\mathbf{X} \setminus \{G\}$, are parametrized by the simple regular R-modules. If S is such a simple regular R-module, then the injective envelope of T_S in $D(R)/\mathcal{T}_{\mathbf{X}}$ is T_P where P denotes the so called Prüfer module corresponding to S. The Prüfer modules have been studied in [**70**]; in particular the structure of their endomorphism rings has been determined. It follows from Ringel's analysis that $S \simeq \operatorname{Ker} \pi_P$ for a generator π_P of $\operatorname{rad} \operatorname{End}_R(P)$. □

In [**57**, 5.5], Lenzing has shown that the functor $C(R)/\mathcal{S}_{\mathbf{X}} \to \operatorname{Presh} \mathbf{X}$, $F \mapsto \operatorname{Hom}_{\mathbf{X}}(T_R, F)$, induces an equivalence between $C(R)/\mathcal{S}_{\mathbf{X}}$ and the category of coherent sheaves on $(\mathbf{X}, \mathcal{O}_{\mathbf{X}})$. This functor extends to an equivalence $D(R)/\mathcal{T}_{\mathbf{X}} \to \operatorname{Qcoh} \mathbf{X}$.

PROPOSITION 14.8. *The functor which sends $F \in D(R)$ to $\operatorname{Hom}_{\mathbf{X}}(T_R, F)$ induces an equivalence between $D(R)/\mathcal{T}_{\mathbf{X}}$ and the category $\operatorname{Qcoh} \mathbf{X}$ of quasi-coherent sheaves on $(\mathbf{X}, \mathcal{O}_{\mathbf{X}})$.*

PROOF. The coherent sheaves form the category of finitely presented objects of the locally coherent category $\operatorname{Qcoh} \mathbf{X}$. Analogously, $\operatorname{fp}(D(R)/\mathcal{T}_{\mathbf{X}}) = C(R)/\mathcal{S}_{\mathbf{X}}$ by Proposition A.5. Therefore the assertion follows from Lenzing's result. □

Composing the equivalence $D(R)/\mathcal{T}_{\mathbf{X}} \to \operatorname{Qcoh} \mathbf{X}$ with the canonical functor $\operatorname{Mod} R \to D(R)/\mathcal{T}_{\mathbf{X}}$, we obtain the following.

COROLLARY 14.9. *The functors*

$$\operatorname{Mod} R \longrightarrow \operatorname{Qcoh} \mathbf{X},\ M \mapsto M_{\mathbf{X}} \quad \textit{and} \quad \operatorname{Qcoh} \mathbf{X} \longrightarrow \operatorname{Mod} R,\ F \mapsto \Gamma(\mathbf{X}, F)$$

induce mutually inverse equivalences between

(1) *the full subcategory of R-modules having no indecomposable preinjective direct summand, and*
(2) *the full subcategory of quasi-coherent sheaves F having a presentation*

$$(\mathcal{O}_{\mathbf{X}})^{(I)} \longrightarrow (\mathcal{O}_{\mathbf{X}})^{(J)} \longrightarrow F \longrightarrow 0.$$

The kernel of $\operatorname{Mod} R \to \operatorname{Qcoh} \mathbf{X}$ *consists of all maps between* R*-modules which factor through a coproduct of indecomposable preinjective* R*-modules.*

PROOF. Combine Theorem 14.4 and Proposition 14.8. $\square$

CHAPTER 15

Coherent rings

In this chapter we investigate the sheaf of definable scalars on the Gabriel spectrum $\operatorname{Sp} R$ of a right coherent ring R. The *Gabriel spectrum* of R is, by definition, the set of isomorphism classes of indecomposable injective R-modules, and the collection of subsets $\mathbf{U}_X = \{M \in \operatorname{Sp} R \mid \operatorname{Hom}_R(X, M) = 0\}$, $X \in \operatorname{mod} R$, forms a basis of open subsets for the (Zariski) topology on $\operatorname{Sp} R$. If R is right coherent, then $\mathbf{X} = \operatorname{Sp} R$ is a Ziegler-closed subset of $\operatorname{Ind} R$. Our aim is to study some of the properties of the structure sheaf $\mathcal{O}_{\mathbf{X}}$. Also, we shall study the functor $\operatorname{Mod} R \to \operatorname{Mod} \mathcal{O}_{\mathbf{X}}$, $M \mapsto \widetilde{M}_{\mathbf{X}}$, and its composition $\operatorname{Mod} R \to \operatorname{Mod} R$, $M \mapsto \Gamma(\mathbf{X}, \widetilde{M}_{\mathbf{X}})$, with the global section functor. If R is commutative noetherian, then $\mathbf{X}$ can be identified with the prime spectrum of R and the functor $\operatorname{Mod} R \to \operatorname{Mod} \mathcal{O}_{\mathbf{X}}$, $M \mapsto \widetilde{M}_{\mathbf{X}}$, induces the usual equivalence between $\operatorname{Mod} R$ and the category $\operatorname{Qcoh} \mathbf{X}$ of quasi-coherent sheaves on $(\mathbf{X}, \mathcal{O}_{\mathbf{X}})$; in particular $\Gamma(\mathbf{X}, \widetilde{M}_{\mathbf{X}}) \simeq M$ for every R-module M. However, if R is non-commutative, then the global section functor $M \mapsto \Gamma(\mathbf{X}, \widetilde{M}_{\mathbf{X}})$ has some interesting properties.

15.1. Preliminaries

In this section we collect some basic facts about coherent rings. Recall that a ring R is *right coherent* if the category $\operatorname{mod} R$ is abelian, equivalently if finitely generated submodules of finitely presented R-modules are finitely presented. Another characterization of right coherent rings is based on fp-injective modules. Recall that an R-module M is *fp-injective* if $\operatorname{Ext}^1_R(X, M) = 0$ for every finitely presented R-module X.

PROPOSITION 15.1. *A ring R is right coherent if and only if the fp-injective R-modules form a definable subcategory of* $\operatorname{Mod} R$.

PROOF. In [**78**] it is shown that R is right coherent if and only if the fp-injective R-modules are closed under forming direct limits. The category of fp-injective R-modules is automatically closed under products and pure submodules. Thus the assertion follows from Theorem 2.1. □

PROPOSITION 15.2. *Let R be a right coherent ring.*

(1) *The injective modules in* $\operatorname{Ind} R$ *form a Ziegler-closed subset* $\mathbf{X}$.
(2) *The functor $C(R) \to \operatorname{mod} R$, $F \mapsto F(R)$, induces an equivalence between $C(R)/\mathcal{S}$ and* $\operatorname{mod} R$ *where $\mathcal{S}$ denotes the Serre subcategory corresponding to* $\mathbf{X}$.
(3) *The functor $D(R) \to \operatorname{Mod} R$, $F \mapsto F(R)$, induces an equivalence between $D(R)/\mathcal{T}$ and* $\operatorname{Mod} R$ *where $\mathcal{T}$ denotes the localizing subcategory of finite type corresponding to* $\mathbf{X}$.

PROOF. (1) This follows from Proposition 15.1 since a module M is injective if and only if M is fp-injective and pure-injective.

(2) This follows from Lemma C.4 since the composition of the equivalence $C(R^{\mathrm{op}})^{\mathrm{op}} \to C(R)$ with $C(R) \to \operatorname{mod} R$ is the right adjoint of the Yoneda functor $\operatorname{mod} R \to C(R^{\mathrm{op}})^{\mathrm{op}}$.

(3) Combine (2) with Proposition A.5. □

From now on we assume for the rest of this section that R is a right coherent ring. We denote by $\mathbf{X} = \operatorname{Sp} R$ the Gabriel spectrum of R; it is the Gabriel spectrum of the locally coherent category $\operatorname{Mod} R$. The basic open sets are of the form $\mathbf{U}_X = \{P \in \mathbf{X} \mid \operatorname{Hom}_R(X, P) = 0\}$ where X is any finitely presented R-module.

LEMMA 15.3. *The topology on the Gabriel spectrum* $\operatorname{Sp} R$ *coincides with the topology which is induced from the Zariski topology on* $\operatorname{Ind} R$.

PROOF. Use Proposition 15.2. □

The definition of the sheaf of definable scalars $\mathcal{O}_{\mathbf{X}}$ and the functor $M \mapsto \widetilde{M_{\mathbf{X}}}$ are based on the fact that $\mathbf{X}$ is a Ziegler-closed subset of $\operatorname{Ind} R$. However, we could also take the direct route via the locally coherent category $\operatorname{Mod} R$ since $\mathbf{X}$ is the Gabriel spectrum of $\operatorname{Mod} R$. More precisely, we have the following.

PROPOSITION 15.4. *The following holds:*

(1) $R_{\mathbf{X}} \simeq \operatorname{End}_{\mathbf{X}}(R)$ *and* $\mathcal{O}_{\mathbf{X}} \simeq \widetilde{\operatorname{End}_{\mathbf{X}}(R)}$.

(2) $M_{\mathbf{X}} \simeq \operatorname{Hom}_{\mathbf{X}}(R, M)$ *and* $\widetilde{M_{\mathbf{X}}} \simeq \widetilde{\operatorname{Hom}_{\mathbf{X}}(R, M)}$ *for every* R*-module* M.

PROOF. Combine the definition of $R_{\mathbf{X}}$ and $M_{\mathbf{X}}$ with Proposition 15.2. □

15.2. Prime spectrum versus Gabriel spectrum

In this section we consider the *prime spectrum* $\operatorname{Spec} R$ of R which is the the set of two-sided prime ideals of R. Given a two-sided ideal $\mathfrak{a}$ let $\mathbf{V}_{\mathfrak{a}} = \{\mathfrak{p} \in \operatorname{Spec} R \mid \mathfrak{a} \subseteq \mathfrak{p}\}$. The collection of subsets $(\mathbf{V}_{\mathfrak{a}})_{\mathfrak{a} \subseteq R}$ is closed under taking finite unions and arbitrary intersections; thus they form the closed sets of a topology on $\operatorname{Spec} R$. Recall that the basic open sets of the Gabriel spectrum $\operatorname{Sp} R$ are of the form $\mathbf{U}_X = \{P \in \operatorname{Sp} R \mid \operatorname{Hom}_R(X, P) = 0\}$ where X is any finitely presented R-module.

A right noetherian ring R is said to be *fully right bounded* provided that the map

$$\alpha\colon \operatorname{Spec} R \longrightarrow \operatorname{Sp} R, \quad \mathfrak{p} \to E(R/\mathfrak{p})$$

which sends a prime ideal $\mathfrak{p}$ to the injective envelope of $R/\mathfrak{p}$ is bijective. For the rest of this section we assume that R is right noetherian and fully bounded.

LEMMA 15.5. *If* $\mathfrak{a}$ *is a right ideal, then there are primes* $\mathfrak{p}_1, \dots, \mathfrak{p}_n$ *in* $\operatorname{Spec} R$ *such that* $\mathbf{U}_{R/\mathfrak{a}} = \mathbf{U}_{R/\mathfrak{q}}$ *for* $\mathfrak{q} = \mathfrak{p}_1 \cdot \dots \cdot \mathfrak{p}_n$.

PROOF. See [**26**]. □

LEMMA 15.6. *If* $\mathfrak{a}$ *is a two-sided ideal, then* $\alpha(\mathbf{V}_{\mathfrak{a}}) = \operatorname{Sp} R \setminus \mathbf{U}_{R/\mathfrak{a}}$.

PROOF. Let $P = E(R/\mathfrak{p})$ with $\mathfrak{p} \in \operatorname{Spec} R$. If $\phi\colon R/\mathfrak{a} \to P$ is a non-zero morphism with $X = \operatorname{Im} \phi$, then $\mathfrak{a} = \operatorname{ann}(R/\mathfrak{a}) \subseteq \operatorname{ann}(X) \subseteq \mathfrak{p}$. Conversely, $\mathfrak{a} \subseteq \mathfrak{p}$ implies that there is a non-zero morphism $R/\mathfrak{a} \to P$. □

LEMMA 15.7. *Let X be a finitely presented module. Then there are primes $\mathfrak{p}_1, \dots, \mathfrak{p}_n$ in* $\operatorname{Spec} R$ *such that* $\mathbf{U}_X = \mathbf{U}_{R/\mathfrak{q}}$ *for* $\mathfrak{q} = \mathfrak{p}_1 \cdot \ldots \cdot \mathfrak{p}_n$.

PROOF. Let $X = \sum_{i=1}^r x_i R$. For each i there is a product $\mathfrak{q}_i$ of prime ideals such that $\mathbf{U}_{x_i R} = \mathbf{U}_{R/\mathfrak{q}_i}$ by Lemma 15.5. Using Lemma 15.6 we obtain $\mathbf{U}_X = \bigcap_{i=1}^r \mathbf{U}_{x_i R} = \mathbf{U}_{R/\mathfrak{q}}$ for $\mathfrak{q} = \mathfrak{q}_1 \cdot \ldots \cdot \mathfrak{q}_r$. □

We have the following consequence.

PROPOSITION 15.8. *Suppose that R is right noetherian and fully bounded. The assignment $\mathfrak{p} \mapsto E(R/\mathfrak{p})$ induces a homeomorphism between the prime spectrum and the Gabriel spectrum of R. In particular, the Gabriel spectrum is a noetherian topological space.*

From now on we assume for the rest of this section that R is commutative and noetherian. We wish to compare the sheaf of rings $\mathcal{O}$ which is defined on $\operatorname{Spec} R$ (see for instance [**34**]) with the (pre)sheaf of definable scalars on $\mathbf{X} = \operatorname{Sp} R$. More precisely, the homeomorphism $\alpha\colon \operatorname{Spec} R \to \operatorname{Sp} R$ induces an equivalence $\alpha_*\colon \operatorname{Sh}\operatorname{Spec} R \to \operatorname{Sh}\mathbf{X}$, and we ask for a relation between $\mathcal{O}_\mathbf{X}$ and $\alpha_*\mathcal{O}$.

THEOREM 15.9. *Let R be a commutative noetherian ring with $\mathbf{X} = \operatorname{Sp} R$.*

(1) *The presheaf of definable scalars $R_\mathbf{X}$ is a sheaf, i.e. $R_\mathbf{X} = \mathcal{O}_\mathbf{X}$. Moreover, $\mathcal{O}_\mathbf{X} \simeq \alpha_*\mathcal{O}$ and therefore $(\operatorname{Spec} R, \mathcal{O})$ and $(\mathbf{X}, \mathcal{O}_\mathbf{X})$ are isomorphic ringed spaces.*
(2) *Let $\mathfrak{p} \in \operatorname{Spec} R$ and denote by P the injective envelope of $R/\mathfrak{p}$. Then R_P is isomorphic to the stalk $\mathcal{O}_\mathfrak{p}$ at $\mathfrak{p}$ and therefore also isomorphic to the localization $R_\mathfrak{p}$ at $\mathfrak{p}$.*

PROOF. (1) The assertions follow from Gabriel's analysis in [**26**, Chap. VI].

(2) The stalk of $R_\mathbf{X}$ at P is isomorphic to the ring R_P of definable scalars for P by Proposition 13.17. The assertion now follows from (1). □

We give an application of the preceding theorem which yields a new interpretation of a continuous 1-parameter family of modules over a k-algebra S if $R = k[T]$. To this end we need to replace the ringed space $(\operatorname{Spec} R, \mathcal{O})$ with the additive space $(\operatorname{Spec} R, \mathcal{O}_{\mathrm{proj}})$. More precisely, we define a sheaf of additive categories $\mathcal{O}_{\mathrm{proj}}$ on $\operatorname{Spec} R$ by $\mathcal{O}_{\mathrm{proj}}(\mathbf{U}) = \operatorname{proj} \mathcal{O}(\mathbf{U})$ for every open $\mathbf{U} \subseteq \operatorname{Spec} R$, where $\operatorname{proj} S$ denotes the category of finitely generated projective S-modules for any ring S. Analogously, the sheaf $(\operatorname{Ind} S, \mathcal{O}_{\operatorname{proj} S})$ is defined for any ring S.

COROLLARY 15.10. *Let R be a commutative noetherian ring and let S be any ring. Suppose there exists an exact functor $f\colon \operatorname{Mod} R \to \operatorname{Mod} S$ which commutes with products and coproducts. Suppose also that $f(M)$ is indecomposable for every indecomposable injective R-module M. Then f induces a morphism $(\operatorname{Spec} R, \mathcal{O}_{\mathrm{proj}}) \to (\operatorname{Ind} S, \mathcal{O}_{\operatorname{proj} S})$ of additive spaces.*

PROOF. Combine Corollary 13.22 with Theorem 15.9. □

We proceed with a discussion of the quasi-coherent sheaves over $(\operatorname{Spec} R, \mathcal{O})$. Given any R-module M, we denote by $\widehat{M}$ the associated sheaf on $\operatorname{Spec} R$.

THEOREM 15.11. *Let R be a commutative noetherian ring with $\mathbf{X} = \operatorname{Sp} R$. Then for every R-module M the associated presheaf $M_\mathbf{X}$ is a sheaf, i.e. $M_\mathbf{X} = \widetilde{M_\mathbf{X}}$. Moreover, there is a functorial isomorphism $\widetilde{M_\mathbf{X}} \simeq \alpha_*\widehat{M}$.*

PROOF. See [**26**, Chap. VI]. □

15.3. The global section functor

Throughout this section R denotes a right coherent ring and $\mathbf{X} = \operatorname{Sp} R$ denotes the Gabriel spectrum of R. Our aim is to study in some examples the sheaf of definable scalars $\mathcal{O}_{\mathbf{X}}$ and the functor $\operatorname{Mod} R \to \operatorname{Mod} \mathcal{O}_{\mathbf{X}}$, $M \mapsto \widetilde{M}_{\mathbf{X}}$. Let us summarize the basic properties of this functor.

THEOREM 15.12. *The functor which sends every R-module M to the associated sheaf $\widetilde{M}_{\mathbf{X}}$ is faithful, left exact, and commutes with direct limits. For every R-module M there is a functorial monomorphism $M \to \Gamma(\mathbf{X}, \widetilde{M}_{\mathbf{X}})$ in* $\operatorname{Mod} R$.

PROOF. Combine Proposition 13.7 and Proposition 15.4. □

We compute now the ring of global sections $\Gamma(\mathbf{X}, \mathcal{O}_{\mathbf{X}})$ in case R is right artinian. We need two lemmas.

LEMMA 15.13. *Suppose that R is right noetherian. If $P \in \operatorname{Sp} R$ is finitely generated as a module over $\operatorname{End}_R(P)^{\mathrm{op}}$, then $R_P \simeq \operatorname{Biend}_R(P)$.*

PROOF. The assertion is an immediate consequence of Corollary 11.10. □

LEMMA 15.14. *Suppose that R is right artinian. Then every P in $\operatorname{Sp} R$ has finite length over $\operatorname{End}_R(P)^{\mathrm{op}}$ and therefore $R_P \simeq \operatorname{Biend}_R(P)$.*

PROOF. A finitely generated submodule of $\operatorname{Hom}_R(R, P) = P$ over $\operatorname{End}_R(P)^{\mathrm{op}}$ is of the form $\operatorname{Hom}_R(R/U, P)$ for some submodule $U \subseteq R$ since P is injective. Therefore P has finite length over $\operatorname{End}_R(P)^{\mathrm{op}}$ since R has finite length over R. The second half of the assertion now follows from the preceding lemma. □

PROPOSITION 15.15. *Suppose that R is right artinian. Then $\Gamma(\mathbf{X}, \mathcal{O}_{\mathbf{X}}) \simeq \prod_{P \in \mathbf{X}} \operatorname{Biend}_R(P)$.*

PROOF. $\operatorname{Sp} R$ is a discrete space and therefore the assertion follows from the preceding lemma and Proposition 13.17 since $\Gamma(\mathbf{X}, \mathcal{O}_{\mathbf{X}})$ is the product of the stalks. □

We exhibit now the case that $\Gamma(\mathbf{X}, \mathcal{O}_{\mathbf{X}})$ is a semi-simple ring. The following lemma will be needed.

LEMMA 15.16. *Let R be right noetherian and let $f\colon \operatorname{Mod} R \to \operatorname{Mod} R$ be a functor having the following properties:*

(1) *there is a functorial monomorphism $\phi_M\colon M \to f(M)$;*
(2) *f sends every monomorphism to a split monomorphism.*

Then the direct limit $f^\infty(M) = \varinjlim_{n\in\mathbb{N}}(f^n(M) \xrightarrow{\phi_{f^n(M)}} f^{n+1}(M))$ is an injective R-module and there is a functorial monomorphism $M \to f^\infty(M)$.

PROOF. The assumption on f implies that ϕ_M factors through the injective envelope $E(M)$ of M. Therefore $f^\infty(M) = \varinjlim E(f^n(M))$ is an injective R-module since a direct limit of injective modules is again injective over a noetherian ring. □

Given an R-module M, we denote by $\Gamma^\infty(\mathbf{X}, \widetilde{M}_{\mathbf{X}})$ the R-module $f^\infty(M)$ where $f(M) = \Gamma(\mathbf{X}, \widetilde{M}_{\mathbf{X}})$.

COROLLARY 15.17. *Suppose that R is right artinian and that $\operatorname{End}_R(P)$ is a division ring for every P in $\mathbf{X}$. Then $\Gamma(\mathbf{X}, \mathcal{O}_{\mathbf{X}})$ is semi-simple and $\Gamma^\infty(\mathbf{X}, \widetilde{M}_{\mathbf{X}})$ is an injective R-module for every R-module M. Therefore every R-module embeds functorially into an injective R-module.*

PROOF. The semi-simplicity of $\Gamma(\mathbf{X}, \mathcal{O}_{\mathbf{X}})$ is a consequence of Proposition 15.15. It follows from Theorem 15.12 that the functor $M \mapsto f(M) = \Gamma(\mathbf{X}, \widetilde{M}_{\mathbf{X}})$ satisfies the conditions (1) - (2) in Lemma 15.16, and therefore $\Gamma^\infty(\mathbf{X}, \widetilde{M}_{\mathbf{X}})$ is injective for every $M \in \operatorname{Mod} R$. □

The example $R = \mathbb{Z}/4\mathbb{Z}$ shows that, in general, a functorial embedding (provided by an additive functor) of R-modules into injective R-modules is impossible.

We turn now our attention to hereditary rings. We begin with some general observations. Let $\mathcal{A} = \operatorname{Mod} R$ and let $\mathcal{S}$ be a Serre subcategory of $\operatorname{mod} R$. Denote by $\mathcal{T} = \varinjlim \mathcal{S}$ the localizing subcategory of $\mathcal{A}$ which is generated by $\mathcal{S}$.

LEMMA 15.18. *The following holds:*

(1) *The section functor $\mathcal{A}/\mathcal{T} \to \mathcal{A}$ commutes with direct limits and induces an equivalence between $\mathcal{A}/\mathcal{T}$ and the perpendicular category $\mathcal{S}^\perp$, i.e. the full subcategory of objects M in $\mathcal{A}$ with $\operatorname{Hom}(\mathcal{S}, M) = 0 = \operatorname{Ext}^1(\mathcal{S}, M)$.*
(2) *If $\operatorname{Ext}^2(\mathcal{S}, -) = 0$, then the inclusion $\mathcal{S}^\perp \to \mathcal{A}$ is exact.*

PROOF. For (1) we refer to Proposition A.6. To prove (2) observe first that the inclusion $\mathcal{S}^\perp \to \mathcal{A}$ is automatically left exact since it has a left adjoint. The assumption on $\mathcal{S}$ implies that for every exact sequence $0 \to L \to M \to N \to 0$ in $\mathcal{A}$ with L and M in $\mathcal{S}^\perp$, also N lies in $\mathcal{S}^\perp$. Therefore the inclusion is also right exact. □

Let $S = \operatorname{End}_{\mathcal{A}/\mathcal{T}}(R)$ and denote by $f\colon R \to S$ the ring homomorphism which is induced by the quotient functor $\mathcal{A} \to \mathcal{A}/\mathcal{T}$.

LEMMA 15.19. *Suppose that $\operatorname{Ext}^2(\mathcal{S}, -) = 0$.*

(1) *$f\colon R \to S$ is a flat epimorphism, i.e. f is an epimorphism of rings and S is flat as an R^{op}-module.*
(2) *$- \otimes_R S\colon \operatorname{Mod} R \to \operatorname{Mod} S$ induces an equivalence $\mathcal{A}/\mathcal{T} \to \operatorname{Mod} S$.*
(3) *Restriction $\operatorname{Mod} S \to \operatorname{Mod} R$ induces an equivalence between $\operatorname{Mod} S$ and $\mathcal{S}^\perp$.*
(4) *f is the universal localization with respect to a collection of maps in $\operatorname{proj} R$.*

PROOF. The quotient functor $\mathcal{A} \to \mathcal{A}/\mathcal{T}$ sends R to a projective generator for $\mathcal{A}/\mathcal{T}$ since the right adjoint is exact by the preceding lemma. Moreover, R is a finitely presented object in $\mathcal{A}/\mathcal{T}$ by Proposition A.5 and therefore $\mathcal{A}/\mathcal{T}$ is equivalent to the category of S-modules; in particular the quotient functor is isomorphic to $- \otimes_R S$. It follows that S is a flat R^{op}-module since the quotient functor is exact. The last assertion follows from Theorem 12.16. □

We can now give a description of the stalks of $\mathcal{O}_{\mathbf{X}}$ and $\widetilde{M}_{\mathbf{X}}$ for every R-module M.

THEOREM 15.20. *Let R be right hereditary and $P \in \mathbf{X}$.*

(1) *$\mathcal{O}_P \simeq R_P$ is again right hereditary and $R \to R_P$ is a flat epimorphism. In fact, R_P is the universal localization with respect to a collection of maps in $\operatorname{proj} R$.*

(2) $(\widetilde{M}_{\mathbf{X}})_P \simeq M \otimes_R R_P$ *for every R-module M.*

PROOF. (1) The isomorphism $\mathcal{O}_P \simeq R_P$ follows from Proposition 13.17, and the properties of R_P follow from Lemma 15.19 by taking the Serre subcategory $\mathcal{S} = \{X \in \operatorname{mod} R \mid \operatorname{Hom}_R(X, P) = 0\}$ in $\operatorname{mod} R$.

(2) Let $\mathcal{T} = \varinjlim \mathcal{S}$. Then

$$(\widetilde{M}_{\mathbf{X}})_P \simeq \operatorname{Hom}_{\operatorname{Mod} R/\mathcal{T}}(R, M) \simeq M \otimes_R R_P$$

where the first isomorphism is taken from Lemma 13.17 and the second is taken from Lemma 15.19. □

We mention some consequences.

COROLLARY 15.21. *Let R be right hereditary. The functor which sends every R-module M to the associated sheaf $\widetilde{M}_{\mathbf{X}}$ is faithful, exact, and preserves coproducts.*

COROLLARY 15.22. *Let R be right hereditary and artinian. Then $\Gamma(\mathbf{X}, \mathcal{O}_{\mathbf{X}})$ is semi-simple and $\Gamma(\mathbf{X}, \widetilde{M}_{\mathbf{X}})$ is an injective R-module for every R-module M. Moreover,*

$$\Gamma(\mathbf{X}, \widetilde{M}_{\mathbf{X}}) \simeq M \otimes_R \Gamma(\mathbf{X}, \mathcal{O}_{\mathbf{X}})$$

and the functor $M \mapsto \Gamma(\mathbf{X}, \widetilde{M}_{\mathbf{X}})$ is exact.

PROOF. For every P in $\operatorname{Sp} R$ the endomorphism ring $\operatorname{End}_R(P)$ is a division ring since R is hereditary. Therefore $\Gamma(\mathbf{X}, \mathcal{O}_{\mathbf{X}})$ is semi-simple by Proposition 15.15. It has already been shown that the functor $M \mapsto \widetilde{M}_{\mathbf{X}}$ is exact and commutes with direct limits. The global section functor $\Gamma(\mathbf{X}, -)\colon \operatorname{Sh} \mathbf{X} \to \operatorname{Ab}$ commutes with direct limits since $\mathbf{X}$ is noetherian, and $\Gamma(\mathbf{X}, -)$ is exact since $\mathbf{X}$ has dimension 0. It follows that $M \mapsto \Gamma(\mathbf{X}, \widetilde{M}_{\mathbf{X}})$ is given by tensoring with $\Gamma(\mathbf{X}, \mathcal{O}_{\mathbf{X}})$. In particular, $\Gamma(\mathbf{X}, \widetilde{M}_{\mathbf{X}})$ is an injective R-module since $\Gamma(\mathbf{X}, \mathcal{O}_{\mathbf{X}})$ is semi-simple and therefore $\Gamma(\mathbf{X}, \widetilde{M}_{\mathbf{X}})$ is injective as an $\Gamma(\mathbf{X}, \mathcal{O}_{\mathbf{X}})$-module. □

EXAMPLE 15.23. Any path algebra over a field of a finite quiver without oriented cycles is hereditary and artinian.

APPENDIX A

Locally coherent Grothendieck categories

A.1. Localization in abelian categories

Throughout this section $\mathcal{A}$ denotes an abelian category. We collect the basic facts about localization in $\mathcal{A}$. For details and proofs we refer to [**26**]. A full subcategory $\mathcal{S}$ of $\mathcal{A}$ is a *Serre subcategory* provided that for every exact sequence $0 \to X' \to X \to X'' \to 0$ in $\mathcal{A}$ the object X belongs to $\mathcal{S}$ if and only if X' and X'' belong to $\mathcal{S}$. The corresponding *quotient category* $\mathcal{A}/\mathcal{S}$ is constructed as follows. The objects of $\mathcal{A}/\mathcal{S}$ are those of $\mathcal{A}$ and

$$\operatorname{Hom}_{\mathcal{A}/\mathcal{S}}(X, Y) = \varinjlim \operatorname{Hom}_{\mathcal{A}}(X', Y/Y')$$

with $X' \subseteq X,\ Y' \subseteq Y$ and $X/X', Y' \in \mathcal{S}$. Again the category $\mathcal{A}/\mathcal{S}$ is abelian and there is canonically defined the *quotient functor* $q\colon \mathcal{A} \to \mathcal{A}/\mathcal{S}$ such that $q(X) = X$; it is exact with $\operatorname{Ker} q = \mathcal{S}$. Here the *kernel* $\operatorname{Ker} f$ of a functor $f\colon \mathcal{A} \to \mathcal{B}$ is, by definition, the full subcategory of all objects X such that $f(X) = 0$. Now suppose that f is exact. Then $\operatorname{Ker} f$ contains $\mathcal{S}$ if and only if f induces a (unique and exact) functor $g\colon \mathcal{A}/\mathcal{S} \to \mathcal{B}$ such that $f = g \circ q$. Note that g is faithful if and only if $\mathcal{S} = \operatorname{Ker} f$.

A Serre subcategory $\mathcal{S}$ of $\mathcal{A}$ is called *localizing* provided that the corresponding quotient functor admits a right adjoint $\mathcal{A}/\mathcal{S} \to \mathcal{A}$. This *section functor* induces an equivalence between $\mathcal{A}/\mathcal{S}$ and the *perpendicular category* $\mathcal{S}^{\perp}$ of objects X satisfying $\operatorname{Hom}(\mathcal{S}, X) = 0 = \operatorname{Ext}^1(\mathcal{S}, X)$. An object in $\mathcal{S}^{\perp}$ is called $\mathcal{S}$*-closed.* Note that an object X in $\mathcal{A}$ is $\mathcal{S}$-closed if and only if the quotient functor $\mathcal{A} \to \mathcal{A}/\mathcal{S}$ induces an isomorphism $\operatorname{Hom}_{\mathcal{A}}(Y, X) \to \operatorname{Hom}_{\mathcal{A}/\mathcal{S}}(Y, X)$ for every Y in $\mathcal{A}$. If $\mathcal{A}$ is a Grothendieck category, then a Serre subcategory $\mathcal{S}$ of $\mathcal{A}$ is localizing if and only if $\mathcal{S}$ is closed under taking coproducts. In this case, $\mathcal{A}/\mathcal{S}$ is again a Grothendieck category.

A.2. Locally coherent Grothendieck categories

Throughout this section we fix a locally coherent Grothendieck category $\mathcal{A}$. Recall that a Grothendieck category $\mathcal{A}$ is *locally coherent* if the full subcategory $\mathcal{C} = \operatorname{fp} \mathcal{A}$ of finitely presented objects is abelian and every object in $\mathcal{A}$ is a direct limit of finitely presented objects. The fp-injective objects of a locally coherent category play an important role. Recall that $M \in \mathcal{A}$ is *fp-injective* if $\operatorname{Ext}^1(X, M) = 0$ for every finitely presented $X \in \mathcal{A}$. The full subcategory of fp-injective objects in $\mathcal{A}$ is denoted by $\operatorname{fpinj} \mathcal{A}$. The following lemma provides a characterization of fp-injectivity, see [**20**].

Lemma A.1. *The following are equivalent for $M \in \mathcal{A}$ and $Z \in \operatorname{fp} \mathcal{A}$.*

(1) $\operatorname{Ext}^1(Z, M) = 0$.

(2) $0 \to \operatorname{Hom}(Z,M) \to \operatorname{Hom}(Y,M) \to \operatorname{Hom}(X,M) \to 0$ *is exact for every exact sequence* $0 \to X \to Y \to Z \to 0$ *in* $\operatorname{fp}\mathcal{A}$.

PROOF. Straightforward. □

We denote by $\operatorname{Lex}(\mathcal{C}^{\mathrm{op}}, \mathrm{Ab})$ the category of left exact functors $\mathcal{C}^{\mathrm{op}} \to \mathrm{Ab}$, and $\operatorname{Ex}(\mathcal{C}^{\mathrm{op}}, \mathrm{Ab})$ denotes the full subcategory of exact functors $\mathcal{C}^{\mathrm{op}} \to \mathrm{Ab}$.

PROPOSITION A.2. *The assignment* $X \mapsto \operatorname{Hom}(-,X)|_{\mathcal{C}}$ *induces equivalences*

$$\mathcal{A} \longrightarrow \operatorname{Lex}(\mathcal{C}^{\mathrm{op}}, \mathrm{Ab}) \quad \textit{and} \quad \operatorname{fpinj}\mathcal{A} \longrightarrow \operatorname{Ex}(\mathcal{C}^{\mathrm{op}}, \mathrm{Ab}).$$

PROOF. For the equivalence $\mathcal{A} \longrightarrow \operatorname{Lex}(\mathcal{C}^{\mathrm{op}}, \mathrm{Ab})$, see [**74**]. The equivalence $\operatorname{fpinj}\mathcal{A} \longrightarrow \operatorname{Ex}(\mathcal{C}^{\mathrm{op}}, \mathrm{Ab})$ then follows from Lemma A.1. □

The *Gabriel spectrum* $\operatorname{Sp}\mathcal{A}$ of $\mathcal{A}$ is the set of isomorphism classes of indecomposable injective objects in $\mathcal{A}$.

PROPOSITION A.3. $\operatorname{Sp}\mathcal{A}$ *cogenerates* $\mathcal{A}$. *Therefore every object in* $\mathcal{A}$ *is subobject of a product* $\prod_{i\in I} M_i$ *with* $M_i \in \operatorname{Sp}\mathcal{A}$ *for every* i.

PROOF. Suppose that X is a non-zero object in $\mathcal{A}$. We need to find a non-zero morphism $X \to M$ for some $M \in \operatorname{Sp}\mathcal{A}$. The object X is a direct limit of finitely presented objects and therefore X has a finitely generated subobject $Y \neq 0$, i.e. Y is a quotient of some finitely presented object. Using Zorn's lemma we find a maximal subobject $Z \subseteq Y$ and we obtain an injective envelope $Y/Z \to M$ for some $M \in \operatorname{Sp}\mathcal{A}$. The composition with $Y \to Y/Z$ then extends to a non-zero morphism $X \to M$.

To embed X in a product of objects in $\operatorname{Sp}\mathcal{A}$ take $I = \bigcup_{M\in \operatorname{Sp}\mathcal{A}} \operatorname{Hom}(X,M)$ and let $\phi\colon X \to \prod_{i\in I} M_i$ with component $\phi_i = i$ for every i. The first part of the assertion implies that ϕ is a monomorphism. □

A.3. Localization in locally coherent Grothendieck categories

Throughout this section $\mathcal{A}$ denotes a locally coherent category with $\mathcal{C} = \operatorname{fp}\mathcal{A}$. We collect the basic facts about localization in $\mathcal{A}$ with respect to certain localizing subcategories. Most of this material is taken from [**50**]. A localizing subcategory $\mathcal{T}$ of $\mathcal{A}$ is said to be of *finite type* provided that the section functor $\mathcal{A}/\mathcal{T} \to \mathcal{A}$ commutes with direct limits. A torsion theory $(\mathcal{T},\mathcal{F})$ for $\mathcal{A}$ is *hereditary* if $\mathcal{T}$ is closed under subobjects and $(\mathcal{T},\mathcal{F})$ is said to be of *finite type* if $\mathcal{F}$ is closed under direct limits.

PROPOSITION A.4. *The following are equivalent for a full subcategory* $\mathcal{T}$ *of* $\mathcal{A}$.

(1) $\mathcal{T}$ *is localizing of finite type.*
(2) $\mathcal{T} = \varinjlim \mathcal{S}$ *for some Serre subcategory* $\mathcal{S}$ *of* $\operatorname{fp}\mathcal{A}$.
(3) *There is a hereditary torsion theory of finite type* $(\mathcal{T},\mathcal{F})$ *for* $\mathcal{A}$.

Moreover, in this case $\mathcal{S} = \mathcal{T} \cap \operatorname{fp}\mathcal{A}$ *and* $\mathcal{F} = \{X \in \mathcal{A} \mid \operatorname{Hom}(\mathcal{S},X) = 0\}$.

PROOF. See [**50**, Theorems 2.6 and 2.8]. □

From now on assume that $\mathcal{T}$ is a localizing subcategory of finite type with $\mathcal{S} = \mathcal{T} \cap \operatorname{fp}\mathcal{A}$ and $\mathcal{F} = \{X \in \mathcal{A} \mid \operatorname{Hom}(\mathcal{S},X) = 0\}$. We obtain the following commutative diagram of exact functors

$$\begin{array}{ccccc} \mathcal{S} & \longrightarrow & \mathcal{C} & \xrightarrow{p} & \mathcal{C}/\mathcal{S} \\ \downarrow{\scriptstyle i'} & & \downarrow{\scriptstyle i} & & \downarrow{\scriptstyle i''} \\ \mathcal{T} & \longrightarrow & \mathcal{A} & \xrightarrow{q} & \mathcal{A}/\mathcal{T} \end{array}$$

where i and i' denote inclusions, p and q denote quotient functors, and i'' is determined by the commutativity of the diagram.

PROPOSITION A.5. *$\mathcal{A}/\mathcal{T}$ is a locally coherent category, and i'' induces an equivalence between $\mathcal{C}/\mathcal{S}$ and the full subcategory of finitely presented objects in $\mathcal{A}/\mathcal{T}$.*

PROOF. It follows from Lemma 1.1 that the quotient functor sends finitely presented objects in $\mathcal{A}$ to finitely presented objects in $\mathcal{A}/\mathcal{T}$. Therefore every object in $\mathcal{A}/\mathcal{T}$ is a direct limit of finitely presented objects. Moreover, $\operatorname{fp}\mathcal{A}/\mathcal{T}$ is abelian since the quotient functor is exact. Thus $\mathcal{A}/\mathcal{T}$ is locally coherent. For the second part of the assertion we refer to [**50**, Theorem 2.6]. □

We list some properties of the section functor $s\colon \mathcal{A}/\mathcal{T} \to \mathcal{A}$.

PROPOSITION A.6. *The section functor induces an equivalence between $\mathcal{A}/\mathcal{T}$ and the perpendicular category $\mathcal{S}^{\perp}$. Therefore $\mathcal{S}^{\perp} = \mathcal{T}^{\perp}$.*

PROOF. See [**50**, Corollary 2.11]. □

PROPOSITION A.7. *The section functor induces an equivalence* $\operatorname{fpinj}\mathcal{A}/\mathcal{T} \to \mathcal{F} \cap \operatorname{fpinj}\mathcal{A}$.

PROOF. In view of the preceding proposition it suffices to show that an object M in $\mathcal{A}/\mathcal{T}$ is fp-injective if and only if $s(M)$ is fp-injective. We apply the characterization of fp-injectivity in Lemma A.1 and use the functorial adjointness isomorphism $\alpha_{X,M}\colon \operatorname{Hom}(X, s(M)) \xrightarrow{\sim} \operatorname{Hom}(q(X), M)$. To prove our claim suppose first that M is fp-injective and let $\varepsilon\colon 0 \to X \to Y \to Z \to 0$ be any exact sequence in $\operatorname{fp}\mathcal{A}$. The image $q(\varepsilon)$ is again an exact sequence of finitely presented objects by Proposition A.5 and therefore an application of α shows that $s(M)$ is fp-injective. Conversely, suppose that $s(M)$ is fp-injective and let $\varepsilon\colon 0 \to X \to Y \to Z \to 0$ be any exact sequence in $\operatorname{fp}\mathcal{A}/\mathcal{T}$. Using Proposition A.5 one finds an exact sequence $\varepsilon'\colon 0 \to X' \to Y' \to Z' \to 0$ in $\operatorname{fp}\mathcal{A}$ such that $q(\varepsilon') \simeq \varepsilon$. Again, an application of α implies that M is fp-injective. □

We denote by $\operatorname{inj}\mathcal{A}$ the full subcategory of injective objects in $\mathcal{A}$.

PROPOSITION A.8. *The section functor induces an equivalence between* $\operatorname{inj}\mathcal{A}/\mathcal{T}$ *and* $\mathcal{F} \cap \operatorname{inj}\mathcal{A}$.

PROOF. The right adjoint of an exact functor sends injective objects to injective objects. The assertion follows since an injective object X belongs to $\mathcal{T}^{\perp}$ if and only if $\operatorname{Hom}(\mathcal{T}, X) = 0$. □

It is interesting to observe that the section functor makes the following diagram of functors commutative

$$\begin{array}{ccc} \mathcal{A}/\mathcal{T} & \xrightarrow{s} & \mathcal{A} \\ \downarrow\wr & & \downarrow\wr \\ \operatorname{Lex}((\mathcal{C}/\mathcal{S})^{\mathrm{op}}, \mathrm{Ab}) & \xrightarrow{p_*} & \operatorname{Lex}(\mathcal{C}^{\mathrm{op}}, \mathrm{Ab}) \end{array}$$

where $p_*(F) = p \circ F$ for every $F \in \operatorname{Lex}((\mathcal{C}/\mathcal{S})^{\mathrm{op}}, \mathrm{Ab})$. Restricting to the fp-injective objects one obtains the following diagram

$$\begin{array}{ccc} \operatorname{fpinj} \mathcal{A}/\mathcal{T} & \longrightarrow & \operatorname{fpinj} \mathcal{A} \\ \downarrow \wr & & \downarrow \wr \\ \operatorname{Ex}((\mathcal{C}/\mathcal{S})^{\mathrm{op}}, \mathrm{Ab}) & \longrightarrow & \operatorname{Ex}(\mathcal{C}^{\mathrm{op}}, \mathrm{Ab}) \end{array}$$

The section functor sends injective objects to injective objects and identifies therefore $\operatorname{Sp} \mathcal{A}/\mathcal{T}$ with a subset $\mathbf{U}$ of $\operatorname{Sp} \mathcal{A}$.

PROPOSITION A.9. *Let X be an object in $\mathcal{A}$.*

(1) *$X \in \mathcal{F}$ if and only if X is a subobject of a product $\prod_{i \in I} M_i$ with $M_i \in \mathbf{U}$ for all i.*
(2) *$X \in \mathcal{T}$ if and only if* $\operatorname{Hom}(X, \mathbf{U}) = 0$.

PROOF. (1) Suppose first that $X \in \mathcal{F}$. Let $X \to M$ be an injective envelope. It is clear that $M \in \mathcal{F}$. Thus $M = s(N)$ for some object $N \in \mathcal{A}/\mathcal{T}$ by Proposition A.8, and we find a monomorphism $N \to \prod_{i \in I} N_i$ in $\mathcal{A}/\mathcal{T}$ with $N_i \in \operatorname{Sp} \mathcal{A}/\mathcal{T}$ for all i by Proposition A.3. Letting $M_i = s(N_i)$ for each i, we obtain a monomorphism $X \to \prod_{i \in I} M_i$ since the section functor is left exact and preserves products. For the converse use that $\mathcal{F}$ is closed under products and subobjects.

(2) Use part (1). □

A.4. Locally noetherian categories

A Grothendieck category $\mathcal{A}$ is said to be *locally noetherian* if there exists a generating set of noetherian objects in $\mathcal{A}$. There are various equivalent conditions and we collect some of them. We need the following analogue of Baer's criterion.

LEMMA A.10. *Let $\mathcal{A}$ be a locally finitely presented Grothendieck category. For an object M in $\mathcal{A}$ the following are equivalent.*

(1) *M is injective, i.e.* $\operatorname{Ext}^1(Z, M) = 0$ *for all $Z \in \mathcal{A}$.*
(2) $\operatorname{Ext}^1(Z, M) = 0$ *for all finitely generated $Z \in \mathcal{A}$.*

PROOF. Suppose that $\operatorname{Ext}^1(Z, M) = 0$ for all finitely generated $Z \in \mathcal{A}$ and let $\phi \colon M \to E$ be an injective envelope. If $\psi \colon Y \to \operatorname{Coker} \phi$ is a morphism with $Y \in \operatorname{fp} \mathcal{A}$, then $Z = Y/\operatorname{Ker} \psi$ is a subobject of $\operatorname{Coker} \phi$. Now ϕ can be written as composition $M \to M \coprod Z \to E$ of two monomorphisms since $\operatorname{Ext}^1(Z, M) = 0$ by assumption. We conclude from the minimality of E that $Z = 0$. Thus $\operatorname{Coker} \phi = 0$ and M is injective. □

PROPOSITION A.11. *Let $\mathcal{A}$ be a locally finitely presented Grothendieck category. Then the following are equivalent:*

(1) *$\mathcal{A}$ is locally noetherian.*
(2) *Every finitely presented object is noetherian.*
(3) *Every fp-injective object is injective.*
(4) *Every direct limit of injective objects is injective.*
(5) *Every injective object is a coproduct of indecomposable objects.*

PROOF. (1) $\Rightarrow$ (2) Clear.

(2) $\Rightarrow$ (3) Finitely generated and finitely presented objects coincide. Therefore Baer's criterion implies that fp-injective objects are injective.

(3) $\Rightarrow$ (4) Clear, since every direct limit of fp-injective objects is again fp-injective.

(4) $\Rightarrow$ (5) Let M be a non-zero injective object and choose a finitely generated non-zero subobject X of M. (4) implies that the union of any chain of injective subobjects of M is again injective. Thus we can use Zorn's lemma and find a maximal direct summand U of M which does not contain X. Let $M = U \coprod V$. Then V is indecomposable since a decomposition of V would contradict the maximality of U. Thus M has an indecomposable direct summand. Using Zorn's lemma we can find a maximal set $(U_i)_{i \in I}$ of indecomposable injective subobjects of M such that the sum $N = \sum_{i \in I} U_i$ is direct. Again, N is a direct summand of M and its complement is zero since otherwise an indecomposable direct summand could be adjoined to the family $(U_i)_{i \in I}$. Therefore M is a coproduct of indecomposable objects.

(5) $\Rightarrow$ (1) Adopt the proof of the classical case $\mathcal{A} = \operatorname{Mod} R$.

$\square$

Note that a decomposition $M = \coprod_{i \in I} M_i$ of an injective object in $\mathcal{A}$ into indecomposable objects is unique up to isomorphism. More precisely, if $M = \coprod_{j \in J} N_j$ is another decomposition into indecomposable objects, then there exists a bijection $\sigma \colon I \to J$ such that $M_i \simeq N_{\sigma(j)}$ for all $i \in I$.

APPENDIX B

Dimensions

B.1. The dimension of an abelian category

Let $\mathcal{C}$ be a skeletally small abelian category. There is a canonical method to assign to $\mathcal{C}$ a dimension which we denote by $\dim \mathcal{C}$, e.g. see [**26, 38**]. Define a filtration $(\mathcal{C}_\alpha)_\alpha$ of Serre subcategories $\mathcal{C}_\alpha \subseteq \mathcal{C}$ recursively as follows:

- $\mathcal{C}_{-1} = 0$;
- if α is an ordinal of the form $\alpha = \beta + 1$, then $\mathcal{C}_\alpha$ consists of all objects in $\mathcal{C}$ which become of finite length in $\mathcal{C}/\mathcal{C}_\beta$;
- if α is a limit ordinal, then $\mathcal{C}_\alpha = \bigcup_{\gamma<\alpha} \mathcal{C}_\alpha$.

Let $\mathcal{C}_\infty = \bigcup_\alpha \mathcal{C}_\alpha$ and denote by $\dim \mathcal{C}$ the smallest ordinal α (or ∞ if such an ordinal does not exist) such that $\mathcal{C}_\alpha = \mathcal{C}$. We list some basic properties of $\dim \mathcal{C}$.

LEMMA B.1. *Let $\mathcal{C}$ and $\mathcal{D}$ be skeletally small abelian categories.*

(1) $\dim \mathcal{C}^{\mathrm{op}} = \dim \mathcal{C}$.
(2) *Let $\mathcal{S}$ be a Serre subcategory of $\mathcal{C}$. Then* $\dim \mathcal{C}/\mathcal{S} \leq \dim \mathcal{C}$.
(3) *Let $f\colon \mathcal{D} \to \mathcal{C}$ be a faithful and exact functor. Then* $\dim \mathcal{D} \leq \dim \mathcal{C}$.
(4) *Let $\mathcal{S}$ be a Serre subcategory of $\mathcal{C}$. Then*

$$\sup(\dim \mathcal{S}, \dim \mathcal{C}/\mathcal{S}) \leq \dim \mathcal{C} \leq \dim \mathcal{S} \oplus \dim \mathcal{C}/\mathcal{S}.$$

PROOF. (1) Clearly, $(\mathcal{C}^{\mathrm{op}})_\alpha = \mathcal{C}_\alpha$, and therefore $\dim \mathcal{C}^{\mathrm{op}} = \dim \mathcal{C}$.

(2) Denote by $q\colon \mathcal{C} \to \mathcal{C}/\mathcal{S}$ the quotient functor. Then $q(\mathcal{C}_\alpha) \subseteq (\mathcal{C}/\mathcal{S})_\alpha$ for every α and therefore $\dim \mathcal{C}/\mathcal{S} \leq \dim \mathcal{C}$.

(3) $f^{-1}(\mathcal{C}_\alpha) \subseteq \mathcal{D}_\alpha$ for every α, and therefore $\dim \mathcal{D} \leq \dim \mathcal{C}$.

(4) $\sup(\dim \mathcal{S}, \dim \mathcal{C}/\mathcal{S}) \leq \dim \mathcal{C}$ follows from (2) and (3). For the other inequality and the definition of $\oplus$ we refer to [**26**, IV, Proposition 1.1]; for instance $\alpha \oplus \beta = \alpha + \beta + 1$ for finite ordinals α and β. □

LEMMA B.2. *Suppose there exists X in $\mathcal{C}$ such that there is no proper Serre subcategory of $\mathcal{C}$ containing X. If* $\dim \mathcal{C} < \infty$, *then* $\dim \mathcal{C} = \alpha + 1$ *for some ordinal* α.

PROOF. Let β be a limit ordinal. If $X \in \mathcal{C}_\beta$, then $X \in \mathcal{C}_\gamma$ for some $\gamma < \beta$, and therefore $\dim \mathcal{C} < \beta$. □

The filtration $(\mathcal{C}_\alpha)_\alpha$ of $\mathcal{C}$ can be used to define for every object X in $\mathcal{C}$ a dimension $\dim X$ which generalizes the Loewy length of a finite length object. If X is of finite length, then $\dim X$ denotes the minimal $n \geq 0$ such that there is a chain

$$0 = X_0 \subseteq X_1 \subseteq \ldots \subseteq X_n = X$$

with X_i/X_{i-1} semi-simple for all i. If X is an arbitrary object, then $\dim X$ denotes the least ordinal of the form $\omega\alpha + n$ (or ∞ if such an ordinal does not exist) such that α is not a limit ordinal and X is of finite length in $\mathcal{C}/\mathcal{C}_{\alpha-1}$ with $\dim X \leq n$. The following property of $\dim X$ is an immediate consequence of the definition.

LEMMA B.3. *Let α be an ordinal. Then $X \in \mathcal{C}_\alpha$ if and only if* $\dim X < \omega(\alpha+1)$.

We list some further properties of $\dim X$.

LEMMA B.4. *Let X, Y be objects in $\mathcal{C}$.*

(1) *If* $0 < \dim X < \infty$, *then* $\dim X = \omega\alpha + n$ *for some non-limit ordinal α and some $n \in \mathbb{N}$.*
(2) *Let $0 \to X' \to X \to X'' \to 0$ be an exact sequence in $\mathcal{C}$. Then*
$$\sup(\dim X', \dim X'') \leq \dim X.$$
If $\dim X' = \omega\alpha' + n'$ *and* $\dim X'' = \omega\alpha'' + n''$, *then* $\dim X \leq \omega\alpha + n$ *for* $\alpha = \sup(\alpha', \alpha'')$ *and* $n = n' + n''$.
(3) $\dim(X \coprod Y) = \sup(\dim X, \dim Y)$.

PROOF. Left to the reader. □

Let $\mathcal{A}$ be a Grothendieck category. Following [**26**], one can assign to $\mathcal{A}$ its *Krull dimension* $\operatorname{Kdim} \mathcal{A}$. Define a filtration $(\mathcal{A}_\alpha)_\alpha$ of localizing subcategories $\mathcal{A}_\alpha \subseteq \mathcal{A}$ recursively as follows:

- $\mathcal{A}_{-1} = 0$;
- if α is an ordinal of the form $\alpha = \beta + 1$, then $\mathcal{A}_\alpha$ is the smallest localizing subcategory of $\mathcal{A}$ containing all objects which become of finite length in $\mathcal{A}/\mathcal{A}_\beta$;
- if α is a limit ordinal, then $\mathcal{A}_\alpha$ is the smallest localizing subcategory of $\mathcal{A}$ containing $\bigcup_{\gamma<\alpha} \mathcal{A}_\alpha$.

The smallest ordinal α (or ∞ if such an ordinal does not exist) such that $\mathcal{A}_\alpha = \mathcal{A}$ is the Krull dimension $\operatorname{Kdim} \mathcal{A}$ of $\mathcal{A}$. There is the following bound for the Krull dimension in case the Grothendieck category is locall coherent.

LEMMA B.5. *Let $\mathcal{A}$ be a locally coherent Grothendieck category. Then the Krull dimension of $\mathcal{A}$ is bounded by* $\dim \operatorname{fp} \mathcal{A}$.

PROOF. Let $\mathcal{C} = \operatorname{fp} \mathcal{A}$. It suffices to show that $\mathcal{C}_\alpha \subseteq \mathcal{A}_\alpha$ for every ordinal α. Clearly this holds for $\alpha = -1$ and for every limit ordinal α provided that $\mathcal{C}_\beta \subseteq \mathcal{A}_\beta$ for every $\beta < \alpha$. Therefore assume $\alpha = \beta + 1$ and $\mathcal{C}_\beta \subseteq \mathcal{A}_\beta$. The quotient functor $\mathcal{A} \to \mathcal{A}/\mathcal{A}_\beta$ induces an exact functor $\mathcal{C}/\mathcal{C}_\beta \to \mathcal{A}/\mathcal{A}_\beta$ which factors through $\operatorname{fp} \mathcal{A}/\mathcal{A}_\beta$ by Proposition A.5. Given $X \in \mathcal{C}$ which is simple in $\mathcal{C}/\mathcal{C}_\beta$, it follows that X is either simple or zero in $\mathcal{A}/\mathcal{A}_\beta$. Therefore $\mathcal{C}_\beta \subseteq \mathcal{A}_\beta$, and this finishes the proof. □

If the Grothendieck category $\mathcal{A}$ is locally noetherian then $\dim \operatorname{fp} \mathcal{A}$ equals the Krull dimension of $\mathcal{A}$ since every localizing subcategory $\mathcal{T}$ of $\mathcal{A}$ is of the form $\mathcal{T} = \varinjlim \mathcal{S}$ for some Serre subcategory $\mathcal{S}$ of $\operatorname{fp} \mathcal{A}$.

B.2. The dimension of a modular lattice

We follow Prest [**61**] and assign to every modular lattice a dimension as follows. Let L be a modular lattice with 0 and 1. Given a pair of elements $x, y \in L$ we define $x \sim y$ if the interval $[x \wedge y, x \vee y]$ has finite length. Note that $\sim$ defines a congruence relation on L. Therefore the set of equivalence classes $L/{\sim}$ carries again the structure of a modular lattice and the canonical map $L \to L/{\sim}$ is a morphism. We define now a cofiltration $(L_\alpha)_\alpha$ of L recursively as follows:

- $L_{-1} = L$;
- if α is an ordinal of the form $\alpha = \beta + 1$, then $L_\alpha = L_\beta/\sim$;
- if α is a limit ordinal, then $L_\alpha = \varinjlim_{\gamma<\alpha} L_\gamma$.

Let $L_\infty = \varinjlim_\alpha L_\alpha$ and denote by $\dim L$ the least ordinal α (or ∞ if such an ordinal does not exist) such that $L_\alpha = 0$. We list some basic properties of $\dim L$.

LEMMA B.6. *Let K and L be modular lattices.*

(1) *Let $K \to L$ be a surjective lattice morphism. Then* $\dim L \leq \dim K$.
(2) *Let $K \to L$ be an injective lattice morphism. Then* $\dim K \leq \dim L$.
(3) $\dim(K \times L) = \sup(\dim K, \dim L)$.

PROOF. Left to the reader. □

LEMMA B.7. *If* $\dim L < \infty$, *then* $\dim L = \alpha + 1$ *for some ordinal* α.

PROOF. Denote for every β by $\pi_\beta \colon L \to L_\beta$ the canonical projection and let $I_\beta = \{x \in L \mid \pi_\beta(x) = 0\}$. If β is a limit ordinal, then $I_\beta = \bigcup_{\gamma<\beta} I_\gamma$. The assertion now follows since $L_\beta = 0$ if and only if $1 \in I_\beta$. □

Recall that a *dense chain* in a lattice L is a sublattice $C \neq 0$ having the property that for every pair $x < y$ in C there is some $z \in C$ with $x < z < y$. Note that having a dense chain is equivalent to having a sublattice isomorphic to $\{m \cdot 2^{-n} \in [0,1] \mid m \in \mathbb{Z}, n \in \mathbb{N}\}$ with its usual ordering. The following lemma is well-known.

LEMMA B.8. $\dim L = \infty$ *if and only if there is a dense chain in L.*

PROOF. Let $\pi \colon L \to L_\infty$ be the canonical map and suppose first $L_\infty \neq 0$. By definition, L_∞ is a dense chain in L_∞. Therefore we can construct inductively a dense chain isomorphic to $\{m \cdot 2^{-n} \in [0,1] \mid m \in \mathbb{Z}, n \in \mathbb{N}\}$ in L, since for any pair $x < y$ in L with $\pi(x) < \pi(y)$ there is some $z \in L$ with $\pi(x) < \pi(z) < \pi(z)$ and therefore $x < z' < y$ for $z' = (x \vee z) \wedge y$. Now suppose that there is a dense chain in L, say between x and y. Using induction one shows that $\pi(x) \neq \pi(y)$ and therefore $\dim L = \infty$. □

Let $\mathcal{C}$ be a skeletally small abelian category. Given an object X in $\mathcal{C}$ we denote by $L(X)$ the lattice of subobjects of X. The dimension of the abelian category $\mathcal{C}$ is closely related to the dimension of the lattices $L(X)$.

LEMMA B.9. *The following are equivalent for every ordinal α:*

(1) $\dim \mathcal{C} \leq \alpha$.
(2) $\dim L(X) \leq \alpha$ *for every object X in $\mathcal{C}$.*

PROOF. The assertion is a consequence of the following elementary fact: Let $\mathcal{S}$ be any Serre subcategory of $\mathcal{C}$. The quotient functor $q \colon \mathcal{C} \to \mathcal{C}/\mathcal{S}$ induces for every X in $\mathcal{C}$ a surjective lattice homomorphism $\pi \colon L(X) \to L(q(X))$ such that $\pi(U) = \pi(V)$ for every pair $U, V \in L(X)$ if and only if $(U+V)/(U \cap V)$ belongs to $\mathcal{S}$. □

The assignment $X \mapsto \dim L(X)$ respects exact sequences in $\mathcal{C}$.

LEMMA B.10. *Let $0 \to X' \to X \to X'' \to 0$ be an exact sequence in $\mathcal{C}$. Then*

$$\dim L(X) = \sup(\dim L(X'), \dim L(X'')).$$

PROOF. Clear. □

APPENDIX C

Finitely presented functors and ideals

C.1. The category of finitely presented functors

Let $\mathcal{C}$ be an additive category. A basic tool for studying the morphisms in $\mathcal{C}$ is the category $\mathrm{fp}(\mathcal{C}, \mathrm{Ab})$ of finitely presented functors $\mathcal{C} \to \mathrm{Ab}$. The first systematic analysis of this category can be found in Auslander's paper on coherent functors [**2**]. Here we collect some of the basic facts about finitely presented functors. In fact, all what is mentioned here is elementary and follows directly from the definitions which are involved.

Given an object X in $\mathcal{C}$ we denote by H_X the representable functor

$$H_X = \mathrm{Hom}(X, -) \colon \mathcal{C} \longrightarrow \mathrm{Ab}, \quad Y \mapsto \mathrm{Hom}(X, Y).$$

Recall that $F \colon \mathcal{C} \to \mathrm{Ab}$ is *finitely presented* if there exists an exact sequence

$$H_Y \longrightarrow H_X \longrightarrow F \longrightarrow 0$$

in $(\mathcal{C}, \mathrm{Ab})$. We recall also Yoneda's lemma.

LEMMA C.1. *For every* $F \colon \mathcal{C} \to \mathrm{Ab}$ *and* X *in* $\mathcal{C}$ *the map* $\mathrm{Hom}(H_X, F) \to F(X)$, $\phi \mapsto \phi_X(\mathrm{id}_X)$, *is an isomorphism which is functorial in* F *and* X.

In particular, the map $\mathrm{Hom}(X, Y) \to \mathrm{Hom}(H_Y, H_X)$, $\phi \mapsto H_\phi = \mathrm{Hom}(\phi, -)$, is bijective for every pair X, Y in $\mathcal{C}$ and therefore the *Yoneda functor*

$$\mathcal{C} \longrightarrow \mathrm{fp}(\mathcal{C}, \mathrm{Ab}), \quad X \mapsto H_X$$

is fully faithful. This shows that every finitely presented functor $F \colon \mathcal{C} \to \mathrm{Ab}$ is of the form $F = \mathrm{Coker}\, H_\phi$ for some map ϕ in $\mathcal{C}$.

LEMMA C.2. *Let* ϕ_1, ϕ_2 *be maps in* $\mathcal{C}$. *Then* $\mathrm{Coker}\, H_{\phi_1} \simeq \mathrm{Coker}\, H_{\phi_2}$ *if and only if there are maps* α_i, β_i, $i = 1, 2$, *in* $\mathcal{C}$ *such that the following diagram commutes*

$$\begin{array}{ccccc} X_1 & \xrightarrow{\alpha_1} & X_2 & \xrightarrow{\alpha_2} & X_1 \\ \downarrow{\scriptstyle \phi_1} & & \downarrow{\scriptstyle \phi_2} & & \downarrow{\scriptstyle \phi_1} \\ Y_1 & \xrightarrow{\beta_1} & Y_2 & \xrightarrow{\beta_2} & Y_1 \end{array}$$

and $\mathrm{id}_{X_1} - \alpha_2 \circ \alpha_1$ *factors through* ϕ_1 *and* $\mathrm{id}_{X_2} - \alpha_1 \circ \alpha_2$ *factors through* ϕ_2.

It is easily checked that $\mathrm{fp}(\mathcal{C}, \mathrm{Ab})$ is an additive category with cokernels. In order to characterize the fact that $\mathrm{fp}(\mathcal{C}, \mathrm{Ab})$ is abelian recall that a map $\psi \colon Y \to Z$ is a *pseudo-cokernel* for $\phi \colon X \to Y$ in $\mathcal{C}$ if the sequence $H_Z \overset{H_\psi}{\to} H_Y \overset{H_\phi}{\to} H_X$ is exact, i.e. every map $\psi' \colon Y \to Z'$ with $\psi' \circ \phi = 0$ factors through ψ.

LEMMA C.3. *The following are equivalent for* $\mathcal{C}$:

(1) $\mathrm{fp}(\mathcal{C}, \mathrm{Ab})$ *is abelian.*
(2) *Every map in* $\mathcal{C}$ *has a pseudo-cokernel.*

In this case an object F in $\operatorname{fp}(\mathcal{C}, \mathrm{Ab})$ is projective if and only if F is a direct summand of H_X for some X in $\mathcal{C}$.

We characterize now the fact that the Yoneda functor has a left adjoint.

LEMMA C.4. *The following are equivalent for $\mathcal{C}$:*

(1) *The Yoneda functor $\mathcal{C} \to \operatorname{fp}(\mathcal{C}, \mathrm{Ab})^{\mathrm{op}}$ has a left adjoint $f\colon \operatorname{fp}(\mathcal{C}, \mathrm{Ab})^{\mathrm{op}} \to \mathcal{C}$.*
(2) *Every map in $\mathcal{C}$ has a cokernel.*

If $\mathcal{C}$ is abelian, then f is exact and induces an equivalence between $\operatorname{fp}(\mathcal{C}, \mathrm{Ab})^{\mathrm{op}}/\mathcal{S}$ and $\mathcal{C}$ where $\mathcal{S}$ denotes the kernel of f.

We consider now extensions of finitely presented functors. To this end fix maps $\phi'\colon X' \to Y'$ and $\phi''\colon X'' \to Y''$ in $\mathcal{C}$ with $F' = \operatorname{Coker} H_{\phi'}$ and $F'' = \operatorname{Coker} H_{\phi''}$. Given a map $\psi\colon X' \to Y''$ let $\phi = \left[\begin{smallmatrix} \phi' & 0 \\ \psi & \phi'' \end{smallmatrix}\right]$ and $F = \operatorname{Coker} H_\phi$. Then we obtain the following commutative diagram with exact rows and columns:

$$\begin{array}{ccccccccc}
0 & \longrightarrow & H_{Y'} & \longrightarrow & H_{Y' \amalg Y''} & \longrightarrow & H_{Y''} & \longrightarrow & 0 \\
 & & \big\downarrow{\scriptstyle H_{\phi'}} & & \big\downarrow{\scriptstyle H_{\phi}} & & \big\downarrow{\scriptstyle H_{\phi''}} & & \\
0 & \longrightarrow & H_{X'} & \longrightarrow & H_{X' \amalg X''} & \longrightarrow & H_{X''} & \longrightarrow & 0 \\
 & & \big\downarrow & & \big\downarrow & & \big\downarrow & & \\
 & & F' & \longrightarrow & F & \longrightarrow & F'' & \longrightarrow & 0 \\
 & & \big\downarrow & & \big\downarrow & & \big\downarrow & & \\
 & & 0 & & 0 & & 0 & &
\end{array}$$

Conversely, any exact sequence $0 \to F' \to F \to F'' \to 0$ gives rise to a commutative diagram of the above form for some map $\psi\colon X' \to Y''$ with $F \simeq \operatorname{Coker} H_\phi$ and $\phi = \left[\begin{smallmatrix} \phi' & 0 \\ \psi & \phi'' \end{smallmatrix}\right]$. Combining these fact one gets the following description of $\operatorname{Ext}^1(F'', F')$ where we assume that ϕ'' has a cokernel map $\pi\colon Y'' \to \operatorname{Coker} \phi''$.

LEMMA C.5. *The following are equivalent for a functor $F\colon \mathcal{C} \to \mathrm{Ab}$:*

(1) *There exists an exact sequence $0 \to F' \to F \to F'' \to 0$.*
(2) *There exists a map $\psi\colon X' \to Y''$ such that $\pi \circ \psi = 0$ and $F \simeq \operatorname{Coker} H_\phi$ for $\phi = \left[\begin{smallmatrix} \phi' & 0 \\ \psi & \phi'' \end{smallmatrix}\right]$*

C.2. Ideals in additive categories

Let $\mathcal{C}$ be an additive category. In this section we collect the basic definitions and some elementary facts about two-sided ideals in $\mathcal{C}$. An *ideal* $\mathfrak{I}$ in $\mathcal{C}$ consists of subgroups $\mathfrak{I}(X, Y)$ in $\operatorname{Hom}(X, Y)$ for every pair of objects X, Y in $\mathcal{C}$ such that for all maps $\alpha\colon X' \to X$ and $\beta\colon Y \to Y'$ in $\mathcal{C}$ the composition $\beta \circ \phi \circ \alpha$ belongs to $\mathfrak{I}(X', Y')$ for every $\phi \in \mathfrak{I}(X, Y)$. Given a functor $F\colon \mathcal{C} \to \mathrm{Ab}$, we denote by $\operatorname{ann} F$ the ideal of maps ϕ in $\mathcal{C}$ satisfying $F(\phi) = 0$. Given a collection Φ of maps and an ideal $\mathfrak{I}$ in $\mathcal{C}$ we define

$$\operatorname{ann} \Phi = \bigcap_{\phi \in \Phi} \operatorname{ann} \operatorname{Coker} H_\phi \quad \text{and} \quad \operatorname{ann}^{-1} \mathfrak{I} = \{\phi \in \mathcal{C} \mid \mathfrak{I} \subseteq \operatorname{ann} \operatorname{Coker} H_\phi\}.$$

Using Yoneda's lemma one obtains the following elementary description of $\operatorname{ann} \Phi$ and $\operatorname{ann}^{-1} \mathfrak{I}$.

LEMMA C.6. *Let $F = \operatorname{Coker} H_\phi$ for some $\phi\colon X \to Y$ in $\mathcal{C}$. Then the following are equivalent for $\psi\colon U \to V$ in $\mathcal{C}$:*

(1) $\psi \in \operatorname{ann} F$, *i.e.* $F(\psi) = 0$.
(2) *The composition* $X \xrightarrow{\alpha} U \xrightarrow{\psi} V$ *factors through* ϕ *for all* $\alpha \in \operatorname{Hom}(X, U)$.

LEMMA C.7. *Let* $\mathfrak{I}$ *be an ideal in* $\mathcal{C}$. *Then the following are equivalent for* $\phi\colon X \to Y$ *in* $\mathcal{C}$:
(1) $\phi \in \operatorname{ann}^{-1}\mathfrak{I}$, *i.e.* $\operatorname{Coker} H_\phi(\mathfrak{I}) = 0$.
(2) *The composition* $X \xrightarrow{\alpha} U \xrightarrow{\psi} V$ *factors through* ϕ *for all* $\psi \in \mathfrak{I}(U, V)$ *and* $\alpha \in \operatorname{Hom}(X, U)$.
(3) *Every map* $X \to V$ *in* $\mathfrak{I}$ *factors through* ϕ.

We discuss now the property of an ideal to be idempotent.

LEMMA C.8. *If* $0 \to F' \to F \to F'' \to 0$ *is exact in* $(\mathcal{C}, \mathrm{Ab})$, *then*

$$(\operatorname{ann} F')(\operatorname{ann} F'') \subseteq \operatorname{ann} F \subseteq (\operatorname{ann} F') \cap (\operatorname{ann} F'').$$

PROOF. Same as for modules. □

LEMMA C.9. *An ideal* $\mathfrak{I}$ *in* $\mathcal{C}$ *is idempotent, i.e.* $\mathfrak{I}^2 = \mathfrak{I}$, *if and only if the class of functors in* $(\mathcal{C}, \mathrm{Ab})$ *vanishing on* $\mathfrak{I}$ *is closed under extensions.*

PROOF. If $\mathfrak{I}^2 = \mathfrak{I}$ then the class of functors vanishing on $\mathfrak{I}$ is closed under extensions by the preceding lemma. To show the converse look at the exact sequence

$$0 \longrightarrow \mathfrak{I}(X, -)/\mathfrak{I}^2(X, -) \longrightarrow \operatorname{Hom}(X, -)/\mathfrak{I}^2(X, -) \longrightarrow \operatorname{Hom}(X, -)/\mathfrak{I}(X, -) \longrightarrow 0$$

for each X in $\mathcal{C}$. □

The preceding lemma explains the following definition. We call an ideal $\mathfrak{I}$ in $\mathcal{C}$ *fp-idempotent* if the class of finitely presented functors in $(\mathcal{C}, \mathrm{Ab})$ vanishing on $\mathfrak{I}$ is closed under extensions. Clearly, any idempotent ideal is fp-idempotent; however the converse is usually not true. Using $\operatorname{ann}^{-1}\mathfrak{I}$ the property of $\mathfrak{I}$ to be an fp-idempotent ideal can be characterized as follows.

LEMMA C.10. *Suppose that the category* $\mathcal{C}$ *has pseudo-cokernels. Then the following are equivalent for an ideal* $\mathfrak{I}$ *in* $\mathcal{C}$:
(1) $\mathfrak{I}$ *is fp-idempotent.*
(2) *If* $\phi', \phi'' \in \operatorname{ann}^{-1}\mathfrak{I}$, *then* $\left[\begin{smallmatrix} \phi' & 0 \\ \psi & \phi'' \end{smallmatrix}\right] \in \operatorname{ann}^{-1}\mathfrak{I}$ *for all* $\psi \in \mathcal{C}$.

PROOF. Let $F' = \operatorname{Coker} H_{\phi'}$ and $F'' = \operatorname{Coker} H_{\phi''}$ be finitely presented functors $\mathcal{C} \to \mathrm{Ab}$. Then there exists an exact sequence $F' \to F \to F'' \to 0$ whenever $F \simeq \operatorname{Coker} H_\phi$ with $\phi = \left[\begin{smallmatrix} \phi' & 0 \\ \psi & \phi'' \end{smallmatrix}\right]$ for some map ψ in $\mathcal{C}$. In fact, it is shown in Lemma C.5 that every exact sequence $0 \to F' \to F \to F'' \to 0$ is of this form.

(1) ⇒ (2) Suppose that $\phi', \phi'' \in \operatorname{ann}^{-1}\mathfrak{I}$ and consider the exact sequence $F' \to F \to F'' \to 0$ corresponding to a map $\psi \in \mathcal{C}$. The image of $F' \to F$ is finitely presented since $\mathcal{C}$ has pseudo-cokernels and is therefore of the form $\operatorname{Coker} H_\alpha$ for some map α which belongs to $\operatorname{ann}^{-1}\mathfrak{I}$. It follows from (1) that $\left[\begin{smallmatrix} \phi' & 0 \\ \psi & \phi'' \end{smallmatrix}\right] \in \operatorname{ann}^{-1}\mathfrak{I}$.

(2) ⇒ (1) Clear. □

Bibliography

[1] J. Asenio Mayor and J. Martinez Hernandez, On flat and projective envelopes, J. Algebra **160** (1993), 434–440.

[2] M. Auslander, Coherent functors, in: Proceedings of the conference on categorical algebra, La Jolla (1966), 189–231.

[3] M. Auslander, Representation theory of artin algebras II, Comm. Algebra **1** (1974), 269–310.

[4] M. Auslander, Large modules over artin algebras, in: Algebra, Topology and Categories, Academic Press (1976), 1–17.

[5] M. Auslander, Functors and morphisms determined by objects, in: Representation theory of algebras. Proc. conf. Philadelphia 1976, ed. R. Gordon, Dekker, New York (1978), 1–244.

[6] M. Auslander and I. Reiten, Representation theory of artin algebras III, Comm. Algebra **3** (1975), 239–294.

[7] M. Auslander and S.O. Smalø, Preprojective modules over artin algebras, J. Algebra **66** (1980), 61–122.

[8] M. Auslander, I. Reiten and S.O. Smalø, Representation theory of artin algebras, Cambridge University Press (1995).

[9] D. Baer, W. Geigle, and H. Lenzing, The preprojective algebra of a tame hereditary Artin algebra, Comm. Algebra **15** (1987), 425–457.

[10] D.J. Benson, J.F. Carlson, and J. Rickard, Complexity and varieties for infinitely generated modules, Math. Proc. Cambridge Philos. Soc. **118** (1995), 223–243.

[11] G.M. Bergman, Coproducts and some universal ring constructions, Trans. Amer. Math. Soc. **23** (1971), 33–87.

[12] K. Burke and M. Prest, Rings of definable scalars and biendomorphism rings, in: Model theory of groups and automorphism groups, London Math. Soc. Lecture Note Ser. **244** (1997), 188–201.

[13] P.M. Cohn, On the free product of associative rings, Math. Z. **71** (1959), 380–398.

[14] W.W. Crawley-Boevey, On tame algebras and bocses, Proc. London Math. Soc. **56** (1988), 451–483.

[15] W.W. Crawley-Boevey, Maps between representations of zero-relation algebras, J. Algebra **126** (1989), 259–263.

[16] W.W. Crawley-Boevey, Regular modules for tame hereditary algebras, Proc. London Math. Soc. **62** (1991), 490–508.

[17] W.W. Crawley-Boevey, Tame algebras and generic modules, Proc. London Math. Soc. **63** (1991), 241–264.

[18] W.W. Crawley-Boevey, Modules of finite length over their endomorphism ring, in: Representations of algebras and related topics, eds. S. Brenner and H. Tachikawa, London Math. Soc. Lec. Note Series **168** (1992), 127–184.

[19] W.W. Crawley-Boevey, Some model theory of modules, Seminar talk at Universität Bielefeld (1993).

[20] W.W. Crawley-Boevey, Locally finitely presented additive categories, Comm. Algebra **22** (1994), 1644–1674.

[21] W.W. Crawley-Boevey, Infinite dimensional modules in the representation theory of finite dimensional algebras, in: I. Reiten, S. Smalø, and O. Solberg (eds.), Algebras and modules I, CMS Conference Proceedings **23** (1998), 29–54.

[22] V. Dlab and C.M. Ringel, Indecomposable representations of graphs and algebras, Mem. Amer. Math. Soc. **173** (1976).

[23] V. DLAB AND C.M. RINGEL, The representations of tame hereditary algebras, in: Representation theory of algebras. Proc. conf. Philadelphia 1976, ed. R. Gordon, Dekker, New York (1978), 329–353.

[24] YU.A. DROZD, Tame and wild matrix problems, Representations and quadratic forms, Institute of Mathematics, Academy of Sciences, Ukrainian SSR, Kiev (1979), 39–74, Amer. Math. Soc. Transl. **128** (1986), 31–55.

[25] E.E. ENOCHS, Injective and flat covers, envelopes and resolvents, Israel J. Math. **39** (1981), 189–209.

[26] P. GABRIEL, Des catégories abéliennes, Bull. Soc. Math. France **90** (1962), 323–448.

[27] P. GABRIEL AND U. OBERST, Spektralkategorien und reguläre Ringe im Von-Neumannschen Sinn, Math. Z. **92** (1966), 389–395.

[28] P. GABRIEL AND J.A. DE LA PEÑA, Quotients of representation finite algebras, Comm. in Algebra **15** (1987), 279–307.

[29] P. GABRIEL AND F. ULMER, Lokal präsentierbare Kategorien, Springer Lecture Notes in Math. **221** (1971).

[30] P. GABRIEL AND M. ZISMAN, Calculus of fractions and homotopy theory, Springer Verlag, Berlin (1967).

[31] S. GARAVAGLIA, Dimension and rank in the theory of modules, East Lansing, preprint (1979).

[32] W. GEIGLE, The Krull-Gabriel dimension of the representation theory of a tame hereditary artin algebra and applications to the structure of exact sequences, Manuscripta Math. **54** (1985), 83–106.

[33] W. GEIGLE AND H. LENZING, Perpendicular categories with applications to representations and sheaves, J. of Algebra **144** (1991), 273–343.

[34] A. GROTHENDIECK AND J. DIEUDONNÉ, Éléments de géometrie algébrique I: Le langage de schémas, Publ. Math. IHES **4** (1960).

[35] A. GROTHENDIECK AND J.L. VERDIER, Prefaisceaux, in: Théorie des Topos et Cohomologie Etale des Schémas, Springer Lecture Notes in Math. **269** (1972), 1–217.

[36] L. GRUSON, Simple coherent functors, in: Representations of algebras, Springer Lecture Notes in Math. **488** (1975), 156–159.

[37] L. GRUSON AND C.U. JENSEN, Deux applications de la notion de L-dimension, C. R. Acad. Sci. Paris Ser. A **282** (1976), 23–24.

[38] L. GRUSON AND C.U. JENSEN, Dimensions cohomologique reliées aux functeurs $\varprojlim^{(i)}$, in: Sém d'Algèbre P. Dubreil et M.-P. Malliavin, Springer Lecture Notes in Math. **867** (1981), 234–249.

[39] I. HERZOG, Elementary duality for modules, Trans. Am. Math. Soc. **340** (1993), 37–69.

[40] I. HERZOG, The Ziegler spectrum of a locally coherent Grothendieck category, Proc. London Math. Soc. **74** (1997) 503–558.

[41] I. HERZOG, The endomorphism ring of a localized coherent functor, J. Algebra **191** (1997), 416–426.

[42] B. HUISGEN-ZIMMERMANN AND S.O. SMALØ, A homological bridge between finite and infinite dimensional representations of algebras, Algebr. Represent. Theory **1** (1998), 169–188.

[43] C.U. JENSEN AND H. LENZING, Model theoretic algebra, Gordon and Breach, New York (1989).

[44] O. KERNER AND A. SKOWROŃSKI, On module categories with nilpotent infinite radical, Compositio Math. **77** (1991), 313–333.

[45] R. KIEŁPIŃSKI, On Γ-pure injective modules, Bull. Acad. Polon. Sci. Sér. Math. Astr. Phys. **15** (1967), 127–131.

[46] G. KRAUSE AND T.H. LENAGAN, Transfinite powers of the Jacobson radical, Comm. Algebra **7** (1979), 1–8.

[47] H. KRAUSE, The endocategory of a module, in: R. Bautista, R. Martínez-Villa, and J.A. de la Peña (eds.), Representation theory of algebras, CMS Conference Proceedings **18** (1996), 419–432.

[48] H. KRAUSE, An axiomatic description of a duality for modules, Adv. Math. **130** (1997), 280–286.

[49] H. KRAUSE, Stable equivalence preserves representation type, Comment. Math. Helv. **72** (1997), 266–284.

[50] H. KRAUSE, The spectrum of a locally coherent category, J. Pure Appl. Algebra **114** (1997), 259–271.

[51] H. KRAUSE, Generic modules over artin algebras, Proc. London Math. Soc. **76** (1998), 276–306.
[52] H. KRAUSE, Exactly definable categories, J. Algebra **201** (1998), 456–492.
[53] H. KRAUSE, Finitistic dimension and Ziegler spectrum, Proc. Amer. Math. Soc. **126** (1998), 983–987.
[54] H. KRAUSE, Functors on locally finitely presented categories, Colloq. Math. **75** (1998), 105–132.
[55] H. KRAUSE AND MANUEL SAORÍN, On minimal approximations of modules, in: E. Green, B. Huisgen-Zimmermann (eds.), Proceedings "Conference on representations of algebras", Seattle 1997, Contemp. Math. **229** (1998), 227–236.
[56] T.S. KUHN, The structure of scientific revolutions, The University of Chicago Press, Chicago (1962).
[57] H. LENZING, Curve singularities arising from the representation theory of tame hereditary artin algebras, in: Representation theory. I. Finite dimensional algebras, Springer Lecture Notes in Math. **1177** (1986), 199–231.
[58] H. LENZING, Auslander's work on artin algebras, in: I. Reiten, S. Smalø, and O. Solberg (eds.), Algebras and modules I, CMS Conference Proceedings **23** (1998), 83–105.
[59] B. MITCHELL, Theory of categories, Academic Press (1965).
[60] J.A. DE LA PEÑA, Constructible functors and the notion of tameness, Comm. Algebra **24** (1996), 1939-1955.
[61] M. PREST, Model theory and modules, London Math. Soc. Lec. Note Series **130** (1988).
[62] M. PREST, Representation embeddings and the Ziegler spectrum, J. Pure Appl. Algebra **113** (1996), 315–323.
[63] M. PREST, The (pre)sheaf of definable scalars, preprint (1995).
[64] M. PREST, A note concerning the existence of many indecomposable infinite dimensional pure-injective modules over some finite dimensional algebras, preprint (1995).
[65] M. PREST, Sheaves of definable scalars, Conference talk at Universität Bielefeld (1996).
[66] M. PREST, Ziegler spectra of tame hereditary algebras, J. Algebra **207** (1998), 146–164.
[67] D. QUILLEN, Higher algebraic K-theory: I, in: Algebraic K-theory I, Springer Lecture Notes in Math. **341** (1973), 85–147.
[68] J. RADA AND M. SAORÍN, Rings characterized by (pre)envelopes and (pre)covers of their modules, Comm. Algebra **26** (1998), 899–912.
[69] J. RICKARD, Idempotent modules in the stable category, J. London Math. Soc. **56** (1997), 149–170.
[70] C.M. RINGEL, Infinite dimensional representations of finite dimensional hereditary algebras, Symp. Math. Ist. Naz. Alta Mat. **23** (1979), 321–412.
[71] C.M. RINGEL, Tame algebras and integral quadratic forms, Springer Lecture Notes in Math. **1099** (1984).
[72] C.M. RINGEL, Recent advances in the representation theory of finite dimensional algebras, in: Representations of finite groups and finite dimensional algebras, Birkhäuser-Verlag, Basel (1991), 141–192.
[73] C.M. RINGEL, The Ziegler spectrum of a tame hereditary algebra, Colloq. Math. **76** (1998), 105–115.
[74] J.-E. ROOS, Locally noetherian categories, in: Category Theory, Homology Theory and their Applications II, Springer Lecture Notes in Math. **92** (1969), 197–277.
[75] A.H. SCHOFIELD, Representations of rings over skew fields, London Math. Soc. Lec. Note Series **92** (1985).
[76] A.H. SCHOFIELD, Universal localization for hereditary rings and quivers, in: Ring theory, Proceedings, Antwerp 1985, Springer Lecture Notes in Math. **1197** (1986), 149–164.
[77] J. SCHRÖER, Hammocks for string algebras, Dissertation, Universität Bielefeld (1997).
[78] B. STENSTRØM, Coherent rings and FP-injective modules, J. London Math. Soc. **2** (1970), 323–329.
[79] B. STENSTRØM, Rings of quotients, Springer-Verlag (1975).
[80] M. ZIEGLER, Model theory of modules, Ann. of Pure and Appl. Logic **26** (1984), 149–213.
[81] B. ZIMMERMANN-HUISGEN, Rings whose right modules are direct sums of indecomposable modules, Proc. Amer. Math. Soc. **77** (1979), 191–197.
[82] B. ZIMMERMANN-HUISGEN, Strong preinjective partitions and representation type of artinian rings, Proc. Amer. Math. Soc. **109** (1979), 309–322.

[83] B. Zimmermann-Huisgen and W. Zimmermann, On the sparsity of representations of rings of pure global dimension zero, Trans. Amer. Math. Soc. **320** (1990), 695–711.
[84] W. Zimmermann, Rein injektive direkte Summen von Moduln, Comm. Algebra **5** (1977), 1083–1117.

Editorial Information

To be published in the *Memoirs*, a paper must be correct, new, nontrivial, and significant. Further, it must be well written and of interest to a substantial number of mathematicians. Piecemeal results, such as an inconclusive step toward an unproved major theorem or a minor variation on a known result, are in general not acceptable for publication. Papers appearing in *Memoirs* are generally longer than those appearing in *Transactions*, which shares the same editorial committee.

As of September 30, 2000, the backlog for this journal was approximately 11 volumes. This estimate is the result of dividing the number of manuscripts for this journal in the Providence office that have not yet gone to the printer on the above date by the average number of monographs per volume over the previous twelve months, reduced by the number of volumes published in four months (the time necessary for preparing a volume for the printer). (There are 6 volumes per year, each containing at least 4 numbers.)

A Consent to Publish and Copyright Agreement is required before a paper will be published in the *Memoirs*. After a paper is accepted for publication, the Providence office will send a Consent to Publish and Copyright Agreement to all authors of the paper. By submitting a paper to the *Memoirs*, authors certify that the results have not been submitted to nor are they under consideration for publication by another journal, conference proceedings, or similar publication.

Information for Authors

Memoirs are printed from camera copy fully prepared by the author. This means that the finished book will look exactly like the copy submitted.

The paper must contain a *descriptive title* and an *abstract* that summarizes the article in language suitable for workers in the general field (algebra, analysis, etc.). The *descriptive title* should be short, but informative; useless or vague phrases such as "some remarks about" or "concerning" should be avoided. The *abstract* should be at least one complete sentence, and at most 300 words. Included with the footnotes to the paper should be the 2000 *Mathematics Subject Classification* representing the primary and secondary subjects of the article. The classifications are accessible from `www.ams.org/msc/`. The list of classifications is also available in print starting with the 1999 annual index of *Mathematical Reviews*. The Mathematics Subject Classification footnote may be followed by a list of *key words and phrases* describing the subject matter of the article and taken from it. Journal abbreviations used in bibliographies are listed in the latest *Mathematical Reviews* annual index. The series abbreviations are also accessible from `www.ams.org/publications/`. To help in preparing and verifying references, the AMS offers MR Lookup, a Reference Tool for Linking, at `www.ams.org/mrlookup/`. When the manuscript is submitted, authors should supply the editor with electronic addresses if available. These will be printed after the postal address at the end of the article.

Electronically prepared manuscripts. The AMS encourages electronically prepared manuscripts, with a strong preference for AMS-LaTeX. To this end, the Society has prepared AMS-LaTeX author packages for each AMS publication. Author packages include instructions for preparing electronic manuscripts, the *AMS Author Handbook*, samples, and a style file that generates the particular design specifications of that publication series. Though AMS-LaTeX is the highly preferred format of TeX, author packages are also available in AMS-TeX.

Authors may retrieve an author package from e-MATH starting from `www.ams.org/tex/` or via FTP to `ftp.ams.org` (login as `anonymous`, enter username as password, and type `cd pub/author-info`). The *AMS Author Handbook* and the *Instruction Manual* are available in PDF format following the author packages link from `www.ams.org/tex/`. The author package can be obtained free of charge by sending email to `pub@ams.org` (Internet) or from the Publication Division, American Mathematical Society, P.O. Box 6248, Providence, RI 02940-6248. When requesting an author package, please specify AMS-LaTeX or AMS-TeX, Macintosh or IBM (3.5) format, and the publication in which your paper will appear. Please be sure to include your complete mailing address.

Sending electronic files. After acceptance, the source file(s) should be sent to the Providence office (this includes any TeX source file, any graphics files, and the DVI or PostScript file).

Before sending the source file, be sure you have proofread your paper carefully. The files you send must be the EXACT files used to generate the proof copy that was accepted for publication. For all publications, authors are required to send a printed copy of their paper, which exactly matches the copy approved for publication, along with any graphics that will appear in the paper.

TeX files may be submitted by email, FTP, or on diskette. The DVI file(s) and PostScript files should be submitted only by FTP or on diskette unless they are encoded properly to submit through email. (DVI files are binary and PostScript files tend to be very large.)

Electronically prepared manuscripts can be sent via email to `pub-submit@ams.org` (Internet). The subject line of the message should include the publication code to identify it as a Memoir. TeX source files, DVI files, and PostScript files can be transferred over the Internet by FTP to the Internet node `e-math.ams.org` (130.44.1.100).

Electronic graphics. Comprehensive instructions on preparing graphics are available at `www.ams.org/jourhtml/graphics.html`. A few of the major requirements are given here.

Submit files for graphics as EPS (Encapsulated PostScript) files. This includes graphics originated via a graphics application as well as scanned photographs or other computer-generated images. If this is not possible, TIFF files are acceptable as long as they can be opened in Adobe Photoshop or Illustrator. No matter what method was used to produce the graphic, it is necessary to provide a paper copy to the AMS.

Authors using graphics packages for the creation of electronic art should also avoid the use of any lines thinner than 0.5 points in width. Many graphics packages allow the user to specify a "hairline" for a very thin line. Hairlines often look acceptable when proofed on a typical laser printer. However, when produced on a high-resolution laser imagesetter, hairlines become nearly invisible and will be lost entirely in the final printing process.

Screens should be set to values between 15% and 85%. Screens which fall outside of this range are too light or too dark to print correctly. Variations of screens within a graphic should be no less than 10%.

Inquiries. Any inquiries concerning a paper that has been accepted for publication should be sent directly to the Electronic Prepress Department, American Mathematical Society, P. O. Box 6248, Providence, RI 02940-6248.

Editors

Selected Titles in This Series

(Continued from the front of this publication)

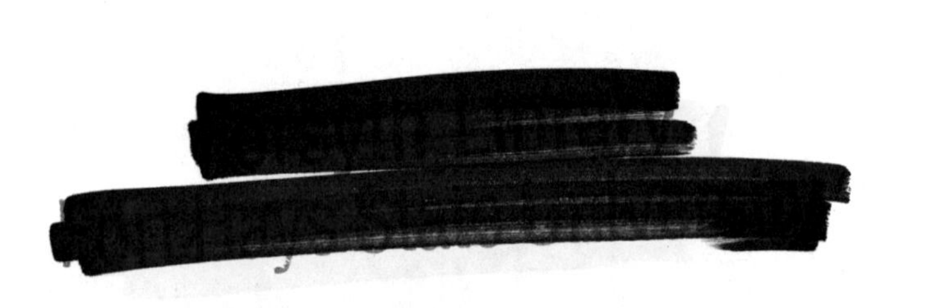